Chaminda S. Bandara
Ranjith Dissanayake

Blast Effects on Reinforced Concrete Cantilevered Slabs

Chaminda S. Bandara
Ranjith Dissanayake

Blast Effects on Reinforced Concrete Cantilevered Slabs

Conventional Design vs Blast Resistant Design

LAP LAMBERT Academic Publishing

Impressum / Imprint

Bibliografische Information der Deutschen Nationalbibliothek: Die Deutsche Nationalbibliothek verzeichnet diese Publikation in der Deutschen Nationalbibliografie; detaillierte bibliografische Daten sind im Internet über http://dnb.d-nb.de abrufbar.

Alle in diesem Buch genannten Marken und Produktnamen unterliegen warenzeichen-, marken- oder patentrechtlichem Schutz bzw. sind Warenzeichen oder eingetragene Warenzeichen der jeweiligen Inhaber. Die Wiedergabe von Marken, Produktnamen, Gebrauchsnamen, Handelsnamen, Warenbezeichnungen u.s.w. in diesem Werk berechtigt auch ohne besondere Kennzeichnung nicht zu der Annahme, dass solche Namen im Sinne der Warenzeichen- und Markenschutzgesetzgebung als frei zu betrachten wären und daher von jedermann benutzt werden dürften.

Bibliographic information published by the Deutsche Nationalbibliothek: The Deutsche Nationalbibliothek lists this publication in the Deutsche Nationalbibliografie; detailed bibliographic data are available in the Internet at http://dnb.d-nb.de.

Any brand names and product names mentioned in this book are subject to trademark, brand or patent protection and are trademarks or registered trademarks of their respective holders. The use of brand names, product names, common names, trade names, product descriptions etc. even without a particular marking in this works is in no way to be construed to mean that such names may be regarded as unrestricted in respect of trademark and brand protection legislation and could thus be used by anyone.

Coverbild / Cover image: www.ingimage.com

Verlag / Publisher:
LAP LAMBERT Academic Publishing
ist ein Imprint der / is a trademark of
OmniScriptum GmbH & Co. KG
Heinrich-Böcking-Str. 6-8, 66121 Saarbrücken, Deutschland / Germany
Email: info@lap-publishing.com

Herstellung: siehe letzte Seite /
Printed at: see last page
ISBN: 978-3-659-57311-8

Zugl. / Approved by: Peradeniya, University of Peradeniya, Diss., 2011

Preface

Studies on blast resistant designs of structures are extremely important in today's world. Knowledge of explosives, blasts, blast wave propagations and the response of structures is essential for structural designers and engineers in designing structural elements against blast loading. This book, a dissertation submitted to the Faculty of Engineering, University of Peradeniya, Sri Lanka in partial fulfillment of the requirements for the degree of Master of the Science of Engineering, is a result of a comprehensive study on blast resistant structural designs.

The reinforced concrete cantilevered slab is one of the most vulnerable structural elements during blast loading. This book focuses mainly on designs of cantilevered slabs. The two main objectives of the study presented in this book are, (i) studying available methodologies for blast load estimations, design of blast resistant reinforced concrete elements and mitigation measures and, (ii) determining the position of conventional cantilevered slabs in blast loading environments using design envelopes, comparing blast resistant designs with conventional designs and thereby identifying improvements needed for conventional cantilever slabs to withstand blast loads.

This book has been written using details from research papers, design codes, text books and our own research. Therefore, we trust that this book will be useful for engineers, designers, researchers and various other groups who are involved in blast resistant designs and constructions.

We would like to thank everyone at the University of Peradeniya who helped to make this book possible; we owe much to Dr. U. I. Dissanayake who assisted in many ways during the research. Thanks also to Dr. P. B. G. Dissanayake for reviewing the work; to Dr. A. L. M. Mauroof, the Coordinator of the Masters Degree Program (2006); and to Professor K. D. W. Nandalal, former head of the Department of Civil Engineering. Special thanks are due to emeritus Professor M. P. Ranaweera for his invaluable advice.

We are grateful to the National Research Council of Sri Lanka, (grant NRC/11-106) for its generous support. We also thank Lambert Academic Publishing, Saarbrücken, Germany for bringing out the book in its present form. Finally, we would like to thank our families for their support and encouragement.

Chaminda S. Bandara

Ranjith Dissanayake

Peradeniya, Sri Lanka

July 2014.

Table of Contents

List of Tables

List of Figures

List of Abbreviations

A	Loaded area
a	Ambient speed of sound in air
a_o	Speed of sound in air
A_s	Tensile reinforcement area within width b
A'_s	Compressive reinforcement area within width b
b	Width of the element
b'	Wave front parameter
C	Coefficient
C_p	Heat capacity at constant pressure
C_v	Heat capacity at constant volume
D	Depth of the element
d_{eff}	Effective depth
E	Modulus of elasticity
E_c	Modulus of elasticity of concrete
E_s	Modulus of elasticity of steel
F	Load, Coefficient
$F(t)$	Load varying with time
f_{cd}	Design strength (static) of concrete
$f_{cd.dyn}$	Dynamic design strength of concrete
$f_{ck.dyn}$	Dynamic characteristic 28 day compressive cylinder strength of concrete
f_{yd}	Design strength (static) of reinforcement
$f_{yd.dyn}$	Dynamic design strength of reinforcement
f_{yk}	Characteristic (static) 28 day compressive cylinder strength of concrete
H	Height of structure
h	Height to a specific location on a structure
H_b	Height of burst
H_m	Height of Mack stem
I	Impulse
I_c	Second moment of area of cracked section
I_s	Incident impulse
I'_s	Incident negative phase impulse

8

I_α	Reflected impulse at angle α
k	Stiffness, Ductility coefficient (static) for reinforcement
K_e	Equivalent elastic stiffness
K_{LM}	Load mass transformation factor
k_{dyn}	Dynamic ductility coefficient for reinforcement
L	Span length of the element
L_x	Length on x direction
M	Mack number
M	Mass
M_p	Ultimate moment capacity
M_{RD}	Design bending moment
P	Pressure
P_o	Ambient pressure
p_d	Dynamic pressure
p_r	Reflected pressure
P_{ro}	Peak reflected over pressure
p_s	Incident pressure
P'_s	Negative pressure
P_{so}	Peak positive incident over pressure
Q_{EXP}	Specific energy of explosive
Q_{TNT}	Specific energy of TNT
R	Standoff distance
R_m	Ultimate resistance of the element
R_o	Limiting distance
R_g	Ground zero distance
r_u	Unit resistance
T	Natural frequency of vibration
t	Time
T_a	Ambient temperature
t_a	Arrival time
t_d	Positive phase duration with liner decay
t_m	Time at which its deflection reaches its maximum
t_n	Negative phase duration

t_s	Positive phase duration
u_m	Mack stem velocity
u_r	Reflected velocity
u_s	Incident velocity
u_s	Wave velocity
u_{sp}	Partial velocity behind blast wave front
W	Equivalent TNT explosive weight
W_{EXP}	Weight of explosive
W_{TNT}	Equivalent TNT charge weight
x	Displacement, deflection
x_e	Equivalent elastic deflection
x_m	Maximum deflection
$x_{m.all}$	Maximum allowable deflection
x_{st}	Displacement if the load is static
Z	Scaled distance
z	Lever arm
α	Angle of incidence
α_{cc}	Coefficient applied to f_{ck} for unfavorable loading effect
β	Angle of blast incident point on the ground from blast point
γ	Specific heat ratio
$\gamma_{c.acc}$	Accidental partial factor for concrete
$\gamma_{s.acc}$	Accidental partial factor for steel
η	Factor on f_{cd} giving magnitude of rectangular stress distribution
θ	Angle of support rotation
λ	Proportion of x over which rectangular stress distribution of magnitude ηf_{cd} acts.
μ	Ductility ratio
ρ	Density of air
ρ_c	Density of concrete
ρ_l	Steel to concrete ratio
ω	Natural circular frequency of vibration
ψ_1, ψ_2	Combination factors

CHAPTER 1

INTRODUCTION

1.1 Importance of studies of blast resistant designs

When designing structures many risks have to be taken in to consideration. Many factors should be incorporated into designs to protect them against damage from excessive wind, floods, water waves, earthquakes, crashing of vehicles or aircraft on to buildings, collapse of masses and explosions. A successful design is one which is practical, economical and safe for occupants and for the structure itself.

Considering the number of terrorist attacks carried out over the past few decades, one of the biggest risks to structures comes from bomb blasts. Bomb attacks on buildings were a common occurrence in Sri Lanka in the 80s and 90s; today they happen even more regularly in other South Asian countries like Afghanistan and Pakistan. As terrorist attacks are becoming increasingly common throughout the world, awareness of blasting effects is important in designing buildings which can withstand the effects of blasts with minimum damage.

A lot of research has been carried out since the 1940s to identify the parameters, behaviour and impact of blast waves on structures. Design methods have been developed to design stronger structures and to improve existing structures to minimize blast effects. Though it not easy to predict where a blast might occur, there are ways to determine the likelihood of a structure to experience blasts. It is useful, therefore, for today's structural designers and engineers to have some knowledge of explosives, blasts, blast wave propagations and response of structures.

1.2 Main areas covered in this book

This book was prepared based on research on the behavior of reinforced concrete cantilever slabs subjected to blast loading. The aims of the research were,

- o Studying available methodologies for blast load estimations, design of blast resistant reinforced concrete elements and mitigation measures.
- o With the aid of the study, determining the position of conventional cantilever concrete slabs in a blast loading environment by developing an envelope to compare with conventional designs and thereby identifying improvements needed for conventional cantilever slabs to withstand blast loads.

Accordingly, this book includes existing methodologies for blast load estimations, design of blast resistant reinforced concrete elements, blast damage mitigation measures, design envelopes for cantilever slabs, comparisons for blast resistant designs with conventional designs and many other important findings of the research.

1.3 Design strategy and codes

Design strategies are determined based on the importance of a structure as well as its risk level. The two main design strategies considered are, (i) designing the whole structure for robustness; this strategy is applicable for buildings with low blast risk, and (ii) element by element design for blast loads; this strategy is applicable for buildings with high blast risk.

Design codes such as BS codes and Eurocodes do not directly provide guidelines for blast resistant designs while UFC standards are especially prepared for blast resistant designs. The research work presented in this book uses BS codes, Eurocodes as well as UFC standards. The important codes, standards and clauses are summarized under two categories;

(i). Design codes that include design methods for robustness and accidental loading but do not include direct provisions for designing against blasts caused by explosives: BS 8110-1:1997, Cl: 3.1.4 for robustness, BS 8110-1:1997, Cl: 3.12.3 for design of ties, BS 8110-2:1985, Cl: 2.6 for robustness, key elements, bridging of elements etc., EN 1990 (2002), expressions for accidental design situations, EN 1991-1-7 (2006) for accidental actions and EN 1992-1-1 (2004), Cl: 2.4.2.4, Cl: 3.1.6, Cl: 9.10 etc., for accidental material factors.

(ii). Design codes which are issued especially for designing against blasts caused by explosives: Unified facilities criteria UFC 4-010-01, 2007, DOD minimum antiterrorism standards for buildings, UFC 3-340-01, 2002, design and analysis of hardened structures to conventional weapons effects, UFC 3-340-02, 2008, structures to resist the effect of accidental explosions.

1.4 Initial considerations to reduce blast risks

Blast risks can be minimized using proper feasibility studies, risk assessments and strong designs. The two main considerations for reducing blast risks on structures are;

(i) Identifying the level of protection required for the structure:
- o Purpose / use of the structure
- o Purpose / use of surrounding structures
- o Security situation of the location

(ii) Deciding on the design strategies:
- o It is important to plan for an effective site layout to limit the possibility of bringing explosives close to the structure by doing the following: Controlling vehicle access, designing parking areas away from the structure, locating trash containers away from the structure, keeping a maximum possible clear space around the structure etc.

- o Structural collapse can be minimized by selecting an effective building layout, using well designed structural elements at easy accessible locations, placing key structural elements at less accessible locations, using of less spans, less number of floors, no cantilever elements, providing sufficient evacuation & emergency access etc.

- o Design methods such as design for robustness, design for alternative load paths to prevent collapse of the structure at an event of a loss of an element, reducing the use of pre-cast & pre-fabricated elements, design for proper structural ductility etc., should be considered.

- o In order to reduce damages caused by flying debris in the event of a blast, the architectural design should contain as little glass, brittle ceilings, brittle partitions etc., as possible.

1.5 Explosions and generation of pressure

Explosion is a very fast reaction due to which a faster and larger expansion occurs that generates huge pressure on air. These pressured air waves travel outwards from the point of explosion as an expanding pressure bulb which grows in size at supersonic velocity.

The pressure bulb is a sphere if the explosion is in the air and a hemisphere if the explosion is on the ground. Because the waves move as either a sphere or a hemisphere, when they pass a structure, different parts of the structure get loaded at different times (t_1, t_2, t_3 etc.,) with varying magnitudes (pressure p_1, p_2, p_3 etc.,) as illustrated in Figure 1.1. Figure 1.2 shows the pressure – time curve at a considered point after a blast.

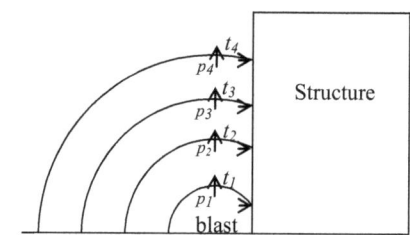

Figure 1.1: Loading of a structure after a blast

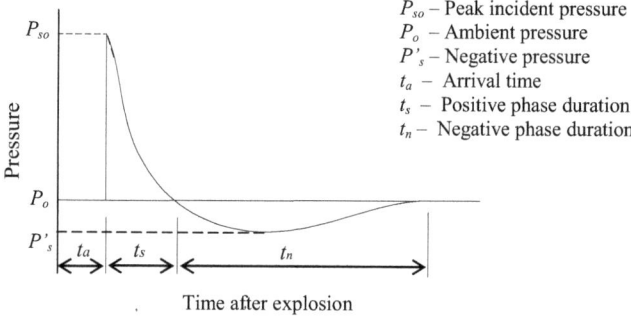

Figure 1.2: Pressure time curve, pressure at a considered point

1.6 Stand-off distance and scaled distance

Figure 1.3 is an illustration showing the stand-off distance (R), ground zero distance (R_g), and angle of incidence (α). The most important parameter is the scaled distance that is,

$$Z = R / W^{1/3} \tag{1.1}$$

where, Z is measured in m/kg$^{1/3}$ and W is the TNT equivalent explosive weight in kg. Hopkinson's Cube Root Law says that,

$$(R_1/R_2) = (W_1/W_2)^{1/3} \qquad (1.2)$$

where, a TNT weight W_1 generates a pressure at a distance R_1 and another TNT weight W_2 generates the same pressure at a distance R_2.

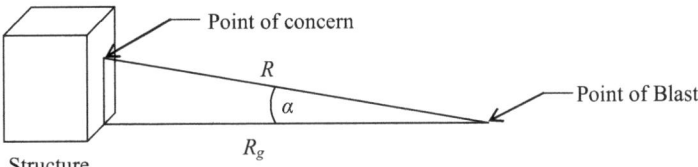

R_g is the ground zero distance, R is the standoff distance and α is the angle of incidence

Figure 1.3: Illustration for Standoff distance and ground zero distance

1.7 Incident and reflected waves

When the pressure wave generated by a blast (incident wave) hits the ground it reflects (reflected wave) and starts moving outwards strengthening the incident wave. Figure 1.4 shows the parameters of both incident and reflected waves.

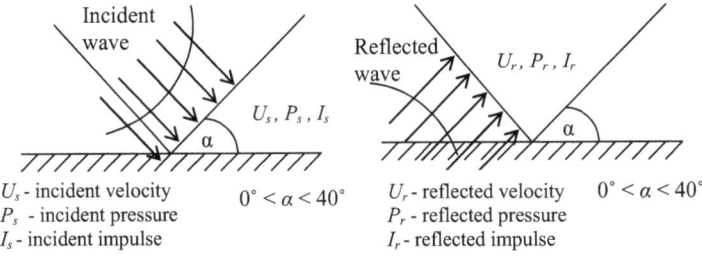

Figure 1.4: Incident and reflected waves

1.8 Regular reflection and Mack reflection

In a surface blast, reflected waves instantly merge with incident waves. In an air blast, depending on the height of blast from the ground, the reflected wave joins the incident wave after a while (i.e., travel time of the incident wave from the blast location to the ground and travel time of the reflected wave). Where the

15

incident angle is greater than 40°, ($\alpha > 40°$), the incident wave reflects on the reflected wave creating an equal pressure region as shown in Figure 1.5.

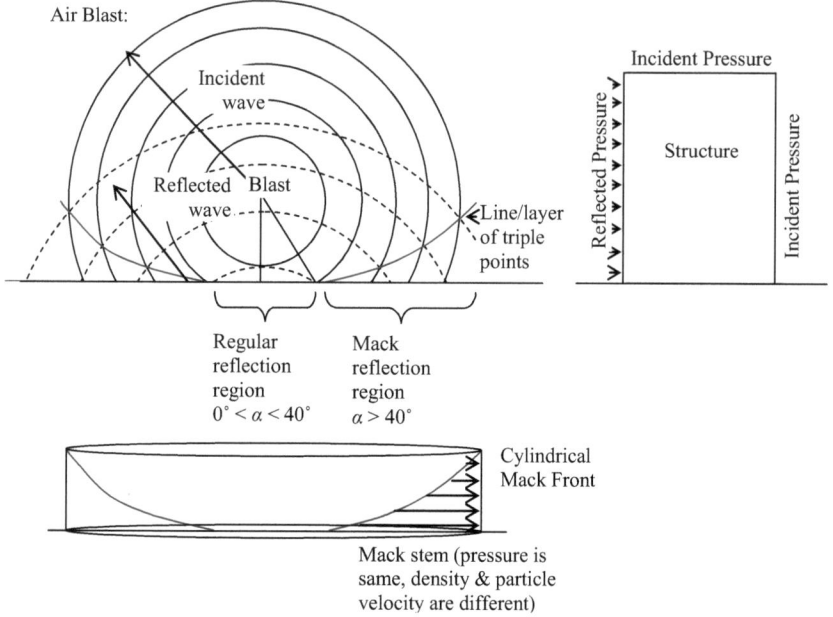

Figure 1.5: Regular reflection and Mack reflection

1.9 Estimation of blast loading parameters

Blast loading parameters are P_s, P_r, U_s, U_r, I_s, I_r, t_s (t_d). Commonly used methods to estimate these parameters are:

(a) Kingery and Bulmash's empirical solutions, (1984)

(b) Ayvazyan et al., (1986), improvements for the coefficient $C_{r\alpha}$ with incident angle and pressure.

The term dynamic pressure describes a different phenomenon and should not be confused with P_s and P_r. Dynamic pressure q_s is a function of kinetic energy of air which is $\frac{1}{2}(\rho u_s^2)$, where, ρ is the density of air and u_s is the particle velocity. Therefore, q_s can be calculated from Eq.(1.3) as required.

$$q_s = 5P_s/\{2(P_s+P_o)\} \qquad (1.3)$$

16

CHAPTER 2

THE DESIGN PHILOSOPHY

2.1 Description

Blast resistant designs differ from conventional designs due to the different characteristics of loading. Blast loads are pulse loads with varying magnitudes and are loading on structures as pulses at different time intervals. In conventional (normal) designs (BS 8110 code series), the loading is mostly static with provisions for live and wind loads (dead loads + live loads). Due to these differences of loading, a knowledge of blast forces and the position (i.e., blast load resistance capacity) of conventionally designed structures in a blast loading environment is important.

Structural elements that are exposed from the main structures such as cantilever slabs are at greater risk in a blast loading environment. Therefore cantilever slabs were selected for the study. Further, the sizes of cantilever slabs selected were the most common and practical sizes (span and effective depth). Steel to concrete ratio was considered up to the limit for which details of previous research work is available. The selected ranges for span, effective depth and steel/concrete ratio are given in Table 2.1.

Table 2.1: Selected range of cantilever slabs for the analysis

Conventional Design			Blast Resistant Design			
Span (mm)	Steel/Concrete Ratio	Effective Depth	Span (mm)	Steel/Concrete Ratio	Effective depth (mm) Impulsive*	QSD**
1,000	As per design		1,000	0.0005 to 0.02	60~400	100~300
1,500	As per design		1,500	0.0005 to 0.02	60~400	100~300
2,000	As per design		2,000	0.0005 to 0.02	60~400	100~300
3,000	As per design		3,000	0.0005 to 0.02	60~400	100~300

*impulsive response of a structure, **quasi static and dynamic response of a structure, refer chapter 2.3 for details*

The spans selected for both conventional design and blast resistant design are from 1,000 mm to 3,000 mm. In blast resistant design, the steel to concrete ratio from 0.0005 to 0.020 was considered for the effective depth range from 60 mm to 400 mm along with a scaled distance range from 0.11 to 38.79 $m/kg^{1/3}$. Since most of the practical designs are within this range the research covers a considerable area of cantilever slab designs.

The magnitudes of loads on the elements for conventional design were selected according to BS 6399-1:1996 and the designs were carried out according to BS 8110-1:1997. Blast load parameters were obtained using Kingery and Bulmash empirical formulae which are given in Table 2.2. The formulae are described below in Chapter 2.2 and more details are given in Annex 1.1.

2.2 Kingery and Bulmash's empirical solutions (1984)

Kingery and Bulmash empirical formulae are widely used to calculate the blast parameters such as pressure, impulse etc. The formulae described in this book are appropriate for hemispherical surface blasts. These equations are available with imperial units and were converted to SI units in the calculations.

(a) Formulae used to determine peak incidence overpressure, P_s (P_{so})

$$T = log\ (Z) \tag{2.1}$$

$$U = -0.7564579301809 + 1.35034249993(T) \tag{2.2}$$

$$Y = 1.9422502013 - 1.6958988741(U) - 0.1541593768146(U^2) +$$
$$0.514060730593(U^3) + 0.0988534365274(U^4) - 0.293912623038(U^5) -$$
$$0.0268112345019(U^6) + 0.109097496421(U^7) +$$
$$0.00162846756311(U^8) - 0.0214631030242(U^9) +$$
$$0.0001456723382(U^{10}) + 0.00167847752266(U^{11}) \tag{2.3}$$

$$P_s = 10^{(Y)} \tag{2.4}$$

where, Z is the scaled distance in ft/lb$^{(1/3)}$. T, U, and Y used in above equations are dummy variables and P_s is given in psi.

(b) Formulae used to determine peak normal reflected overpressure, P_r (P_{ro})

$$A = -0.789312405513 + 1.36637719229(T) \tag{2.5}$$

$$C = 2.56431321138 - 2.21030870597(A) - 0.218536586295(A^2) +$$
$$0.895319589372(A^3) + 0.24989009775(A^4) - 0.569249436807(A^5) -$$
$$0.11791682383(A^6) + 0.224131161411(A^7) + 0.0245620259375(A^8) -$$
$$0.045511600269(A^9) - 0.00190930738887(A^{10}) +$$
$$0.00361471193389(A^{11}) \tag{2.6}$$

$$P_r = 10^{(C)} \tag{2.7}$$

where, A and C are dummy variables. P_r is given in psi.

(c) Formula used to determine incidence impulse, I_s

For $Z \leq 2.41$

$$V=0.832468843425+3.07603296666(T) \tag{2.8}$$

$$I=1.57159240621-0.502992763686(V)+0.171335645235(V^2)+ \\ 0.045017696305(V^3)-0.0118964626402(V^4) \tag{2.9}$$

For $Z > 2.41$

$$V=-2.91358616806+2.40697745406(T) \tag{2.10}$$

$$I=0.71985265584-0.384519026965(V)-0.0260816706301(V^2)+ \\ 0.00595798753822(V^3)+0.014544526107(V^4)- \\ 0.00663289334734(V^5)-0.00284189327204(V^6)+ \\ 0.0013644816227(V^7) \tag{2.11}$$

$$I_s=10^{(I)}\ W^{1/3} \tag{2.12}$$

where, V is a dummy variable and I_s is given in psi ms.

(d) Formulae to determine normal reflected impulse, I_r

$$E=-0.781951689212+1.33422049854(T) \tag{2.13}$$

$$F=1.75291677799-0.949516092853(E)+0.112136118689(E^2)- \\ 0.0250659183287(E^3) \tag{2.14}$$

$$I_r=10^{(F)}\ W^{1/3} \tag{2.15}$$

where, E and F are dummy variables. I_r is given in psi ms.

(e) Formulae to determine wave front velocity, U

$$G=-0.755684472698+1.37784223635(T) \tag{2.16}$$

$$H=0.44977431-0.698029763(G)+0.15891679(G^2)+0..443812098(G^3)- \\ 0.1134020239(G^4)-0.3698870751(G^5)+0.1292305675(G^6)+ \\ 0.198579812(G^7)-0.08676362174(G^8)-0.06203919002(G^9)+ \\ 0.03074829266(G^{10})+0.01026572344(G^{11})-0.0054653325(G^{12})- \\ 0.000693181(G^{13})+0.0003847495(G^{14}) \tag{2.17}$$

$$U=10^H \tag{2.18}$$

where, H and G are dummy variables and U is given in ft/ms.

(f) Formulae to determine positive phase duration, t_s

For $0.45 < Z < 2.54$

$$S=-0.1790217052+5.25099193925(T) \tag{2.19}$$

$$B=0.728671776005+0.130143717675(S)+0.134872511954(S^2)+ \\ 0.0391574276906(S^3)+0.00\ 475933664702(S^4)- \\ 0.0042888144598008(S^5) \tag{2.20}$$

For $2.54 < Z < 7$

$$S=-5.85909812338+9.2996288611(T) \tag{2.21}$$

$$B=0.2009657334-0.0297944268976(S)+0.030632954288(S^2)+ \\ 0.0183405574086(S^3)-0.0173964666211(S^4)- \\ 0.00106321963633(S^5)+0.0056206003097736(S^6) \\ +0.0001618217499(S^7)-0.000686018944(S^8) \tag{2.22}$$

For $Z > 7$

$$S=-4.92699491141+3.46349745571(T) \tag{2.23}$$

$$B=0.5724624769964+0.0933035304009(S)- \\ 0.0004849420883(S^2)-0.00226884995013(S^3)- \\ 0.00295908591505(S^4)+0.00148029868929(S^5) \tag{2.24}$$

$$t_s=10^{(B)}\ W^{1/3} \tag{2.25}$$

where, S is a dummy variable and t_s is given in ms.

These formulae are valid for the region 0.164 ft/lb$^{1/3} \leq Z \leq 100$ ft/lg$^{1/3}$ (0.067 m/kg$^{1/3} \leq Z \leq 40$ m/kg$^{1/3}$ in SI units). Blast wave front parameters produced in a hemispherical surface burst can be calculated by the above empirical formulae having Z as the key parameter.

As the reflection depends on the angle of incidence α, the reflection coefficient at the angle α which is $C_{r\alpha}$ vs α has been plotted after experiments published by Ayvazyan et al. (1986). An illustration for the variation of P_s against α is enclosed in Annex 1.2. Considering the effect of α on I_r, using the equation, $I_\alpha = I_s\ (1+Cos\alpha - 2Cos^2\alpha) + I_rCos^2\alpha$, a modified value for the impulse I_α can be calculated.

With varying α, the regular reflection and Mack reflection regions change. Therefore, blast loading on part of the structure can be due to regular reflected waves while on the other part due to Mack reflected waves; a detailed analysis of this is necessary for a safe design. In this research, it was considered that pressure acts normal to cantilever surface from bottom and hence $\alpha = 0^\circ$ (surface blast, regular reflection).

Table 2.2: Blast load parameters using Kingerly and Bulmash formulae

Z (m/kg$^{1/3}$)	P_r (kPa)	I_r (kPa ms)	t_s (ms)
0.11	426,905.83	154,297.35	
0.22	165,803.32	42,879.59	
0.43	51,997.75	13,799.85	1.13
0.65	23,489.60	7,542.31	2.11
0.86	12,001.87	5,029.04	5.56
1.08	6,644.99	3,717.39	9.29
1.29	3,942.38	2,925.36	10.28
1.51	2,488.13	2,400.41	10.00
1.72	1,658.67	2,029.27	9.55
1.94	1,159.92	1,754.13	9.47
2.15	845.49	1,542.63	9.70
2.37	638.76	1,375.34	10.20
2.59	497.69	1,239.93	11.11
2.80	398.21	1,128.21	12.36
3.02	326.01	1,034.55	13.21
3.23	272.23	954.96	13.94
3.45	231.27	886.53	14.58
3.66	199.42	827.09	15.14
3.88	174.21	775.01	15.64
4.09	153.93	729.01	16.09
4.31	137.37	688.09	16.49
4.53	123.67	651.47	16.86
4.74	112.21	618.51	17.20
4.96	102.51	588.69	17.51
5.17	94.22	561.58	17.81
5.39	87.08	536.84	18.08
6.46	62.57	439.74	19.28
7.54	48.44	372.23	20.28
8.62	39.33	322.59	21.18
10.77	28.35	254.53	22.76
12.93	22.00	210.05	24.14
15.08	17.89	178.71	25.34
17.24	15.03	155.43	26.39
19.39	12.91	137.44	27.31
21.55	11.27	123.13	28.13
23.70	9.96	111.46	28.87
25.86	8.88	101.77	29.53
28.01	7.97	93.59	30.15
30.17	7.18	86.60	30.73
32.32	6.51	80.54	31.28
38.79	4.98	66.45	32.90

2.3 Blast resistant design philosophy

Blast resistant cantilever designs described in this book were carried out according to the procedures described by Cormie et al., (2009), McCann and Smith (2007), Agnew et al., (2007), UFC-3-340-02 (2008), BSEN 1990:2002, BSEN 1992 code series, BS 8110-1:1997 and BS 8110-2:1985. The design philosophy is described below.

Considering the link between the duration of loading of blast pressure on a structure and the natural frequency of the structure, the response of the structure to blast loading can be determined. According to these, three types of response regimes are identified, (i) quasi static, (ii) impulsive and (iii) dynamic.

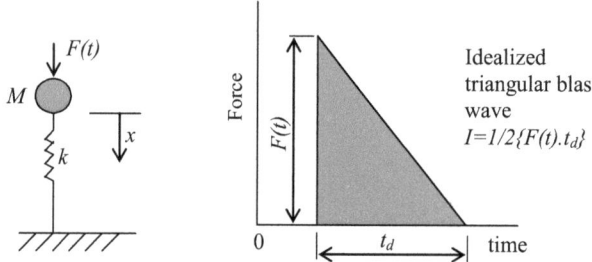

Figure 2.1: Blast force $F(t)$ acting on elastic, single degree of freedom structure

The equation of load pulse $F(t)$ is given by,

$$F(t) = F(1-t/t_d) \qquad (2.26)$$

The equation of motion without damping is given by,

$$M\ddot{x} + kx = F(1-t/t_d) \qquad (2.27)$$

Solution for positive phase ($0 < t < t_d$) is,

$$x(t) = (F/k)(1-Cos\omega t) + (F/(kt_d))(Sin\omega t/\omega - t) \qquad (2.28)$$

where, natural frequency of the element, $\omega = (k/M)^{1/2}$ and k is the stiffness.

The maximum dynamic displacement x_{max} occurs when the velocity,

$d(x(t))/dt = 0$ and then,

$$0 = \omega Sin(\omega t_m) + (1/t_d)(Cos(\omega t_m)) - 1/t_d \qquad (2.29)$$

where, t_m is the time at which displacement reaches its maximum, x_{max}.

Solutions can be presented in the general form,

$\omega t_m = f(\omega t_d)$ and $x_{max}/(F/k) = \psi(\omega t_d) = \psi'(t_d/T)$ (2.30)

where, T is the natural period of vibration of the structure (element).

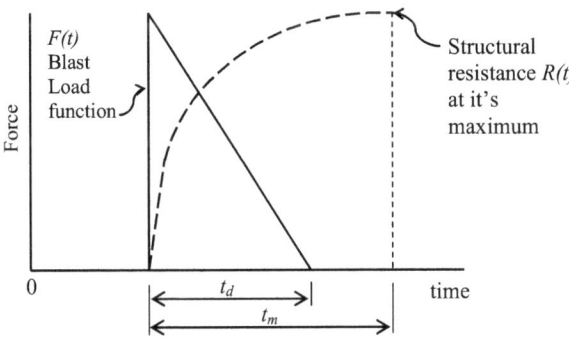

Figure 2.2: Blast load function and structural resistant function

The response of a structure is quasi-static if $10T < t_d$ and $t_m < 0.3t_d$
The response of a structure is impulsive if $t_d < 0.1T$ and $3t_d < t_m$
The response of a structure is dynamic if $0.1T < t_d < 10T$ and $0.3t_d < t_m < 3t_d$.

For quasi static response regime, the work done by the blast load on a structure is equal to the strain energy acquired by the structure. Therefore,

$F x_{max} \qquad = \frac{1}{2} k^2 x_{max}$ (2.31)

$x_{max}/(F/k) \quad = 2 x_{max}/k$ (2.32)

For static load,

$F \qquad = k x_{st}$ (2.33)

Therefore, $x_{max}/(F/k)$ can be given by,

$x_{max}/(F/k) = x_{max}/x_{st} = 2$ (2.34)

where, x_{st} is the displacement if F is a static load.

$x_{max}/(F/k)$ is the dynamic load factor (DLF) that gives the upper bound response of the structure (the quasi static asymptote). In other words, the dynamic deflection of the structure can be obtained by multiplying the static deflection produced by the load applied statically by DLF. The DLF will be above 1.0 due to the increase of effective blast load and may be below 1.0 when decreasing the effective loading (shock amortization).

For impulsive response regime, kinetic energy delivered to the structure is equal to the strain energy acquired by the structure. Therefore,

$$\tfrac{1}{2}\,(I^2/M) \quad = \tfrac{1}{2}\,k^2\,x_{max} \tag{2.35}$$

Then substituting and rearranging the equation,

$$x_{max}/(F/k) \quad = x_{max}/x_{st} = \tfrac{1}{2}\,\omega t_d \tag{2.36}$$

This equation is known as the impulsive asymptote of response.

Accordingly, as presented by Cormie et al., (2009), the structural response curve that is the plot of $x_{max}/(F/k)$ vs t_d/T, is shown in Figure 2.3.

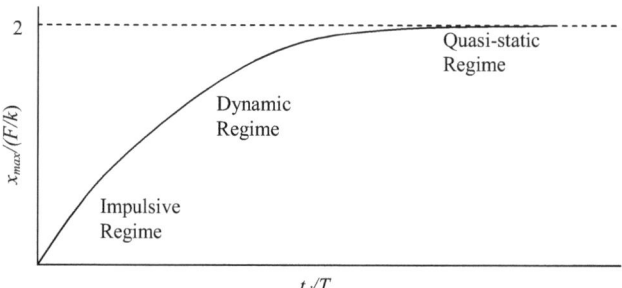

Figure 2.3: Structural response curve

2.4 Parameters for blast resistant design

Blast loading should be incorporated into designs depending on the importance of the structure. Fundamental requirements for the reliability of construction work are given in Eurocodes considering events such as explosions, impact and human errors. Appropriate partial load factors are described in Eurocodes and are simultaneously used with a combination of wind and imposed loads. The design should be based on ultimate limit state with the assumption that there will be no repetitive blast loads on an element (design for one occurrence of blast).

The material strength should be nominal strength. However due to rapid loading, the mechanical properties of material change. Therefore the static strengths of materials have to be converted to dynamic strengths. The factors used for this are called dynamic increase factors (DIF) and are given in Table 2.3. However the modulus of elasticity of both steel and concrete remain unchanged.

Table 2.3: Typical dynamic increase factors for reinforced concrete

Description of stress	Factor	Bending	Shear	Compression
Concrete	$f_{ck.dyn}/f_{ck}$	1.25	1.00	1.15
Steel reinforcement	$f_{yk.dyn}/f_{yk}$	1.20	1.10	-
	$(k_{dyn}f_{yk.dyn})/(kf_{yk})$		1.10	-

According to EN 1992-1-1 (2004), Cl: 2.4.2.4 & Table 2.1N, Accidental material factors should be applied on design strengths of materials. Recommended values; for $\gamma_{c.acc}$ is 1.2 (for concrete) and for $\gamma_{s.acc}$ is 1.0 (for reinforcement). According to EN 1992-1-1 (2004), Cl: 3.1.2 ~ 3.1.6, the coefficient α_{cc} which is applied to concrete strength to take unfavorable loading effect should be 0.85 for bending and compression and, 1.00 for shear. With the use of these partial factors, the dynamic design strengths that have to be used for blast resistant designs are mentioned in Table 2.4.

Table 2.4: Dynamic design strengths for reinforced concrete

Description of stress	Protection Category	Concrete $f_{cd.dyn}=(\alpha_{cc}f_{ck.dyn})/(\gamma_{c.acc})$	Reinforcement $f_{yd.dyn}$
Bending	1	$f_{cd.dyn}= 0.89f_{ck}$	$f_{yd.dyn}= 1.2f_{yk}$
	2	$f_{cd.dyn}= 0.89f_{ck}$	If $k \leq 1.15$, $k_{dyn} \leq 1.0$ $f_{yd.dyn}= 1.2f_{yk}$ If $k \geq 1.15$, $kdyn \geq 1.0$ $f_{yd.dyn}=f_{yk.dyn}/(\gamma s_{acc})$
Shear	1 and 2	$f_{cd.dyn}= 0.83f_{ck}$	$f_{yd.dyn}= 1.1f_{yk}$
Compression	1 and 2	$f_{cd.dyn}= 0.81f_{ck}$	$f_{yd.dyn}= 1.1f_{yk}$

k_{dyn} in the above table for $f_{yd.dyn}$ for reinforcement in bending for protection category 2 is, $k_{dyn} = 0.875k$.

The design compressive strength of concrete, $f_{cd.dyn}$ is derived from characteristic compressive cylinder strength, $f_{ck.dyn}$.

Deformation or deflection limit is the controlling criteria of elements for design for blast loads. The two methods to specify the deformation are support rotation

(θ) and ductility ratio μ, where μ = (total deflection) / (deflection at elastic limit) = x_m/x_e (Figure 2.4).

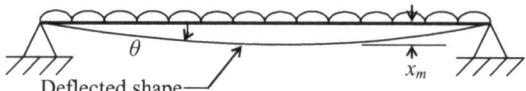

Figure 2.4: Deflection of a simply supported beam, slab or panel

In general, support rotation is used to express deformation in reinforced concrete elements. Based on the magnitude of support rotation, protection categories have been recommended for reinforced concrete beams and slabs. The support rotation, θ = 2° comes under protection category 1 which protects people and equipment from blast pressure. Structure will act as a shield to cover fragments and falling portions of the structure. Protection category 2 is recommended for the protection of the structural elements from collapse under blast loads and value of the support rotation, θ = 4°. As these limits imply plastic deformations of the elements, after a blast, the elements should be repaired or replaced. Shear reinforcement should be provided for protection category 1, for θ > 1° (BS 8666:2000). Under protection category 2, the support rotations up to θ = 8° is allowed provided that the elements have sufficient lateral restraint to develop tensile membrane action.

For θ in the range 0° to 2°, concrete is effective in resisting moment and concrete cover on both surfaces of the element remains uncrushed (type 1 section in Figure 2.5). The design resistant moment M_{Rd} of type 1 section may be determined using conventional plastic theory based upon dynamic design strengths of concrete $f_{cd,dyn}$ and reinforcement $f_{yd,dyn}$. Annex 1.3 shows the equivalent SDOF factors for cantilever slabs.

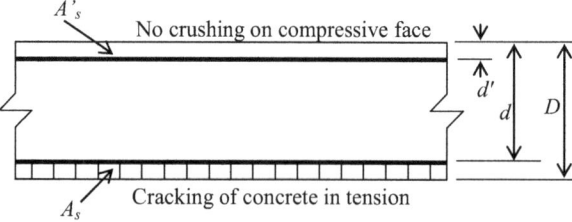

Figure 2.5: Reinforced concrete type 1 section

For $\theta \geq 2°$, concrete gets crushed in compression and hence in the absence of compression reinforcement, the structure may fail. Therefore symmetrical reinforcement at tension and compression faces is recommended. Elements which sustain crushing of concrete without any disengagement of cover at the tensile face are known as type 2 sections (Figure 2.6) where θ remains between 2° and 5°. In this type of section, concrete does not contribute structurally but contribute to the mass and therefore to the inertial resistance. The design resistance moment of type 2 section may be determined by the equation,

$$M_{Rd} = (A_s f_{yd,dyn} z)/b \tag{2.37}$$

where, z is the lever arm of centroids of compression and tensile reinforcements.

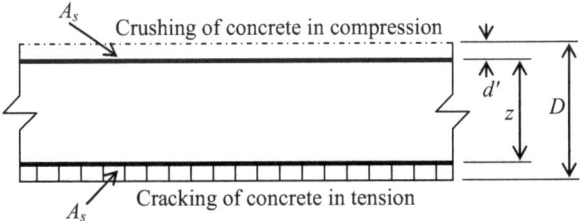

Figure 2.6: Reinforced concrete type 2 section

The recommended idealized resistance – deflection function is shown in Figure 2.7. The ultimate dynamic resistance R_m can be determined using plastic theory and to be modified for static loads at the time of blast.

For $\theta < 5°$, the elastic part of idealized resistance – deflection function is described by Eq. (2.38),

$$I^2 A^2/(2K_{LM}M) = R_m x_e/2 + R_m/(x_m-x_e) \tag{2.38}$$

where, A is the loaded area and K_{LM} is the load mass transformation factor (Annex 1.3 for equivalent SDOF factors for cantilevers with uniformly distributed loads).

Blast loads can be idealized to traingular pressure – time functions as shown in Figure 2.8. The loading on a structure can be assumed as uniformly distributed and represented by specific impulse I.

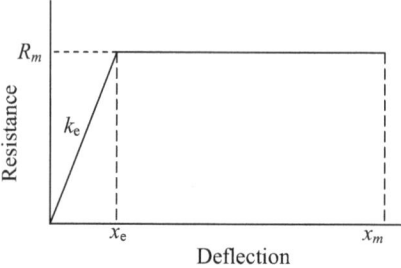

Figure 2.7: Resistance – deflection function (Idealized)

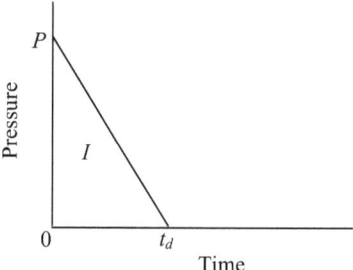

Figure 2.8: Blast pressure – time function (Idealized)

2.5 Steps of design

The design sequence described below is the method described by Cormie et al., (2009). One of the other methods is to use BSEN 1990:2002, NA2.2.5 with tables NA.A1.1, NA.A1.3 which are established to determine the combination factors for imposed loads and wind loads (ψ_1 and ψ_2). Since BS 8110 series of codes are the most common codes for designing reinforced concrete elements in most of the regions in the world, BS 8110-1:1997 is used for the design calculations. Then the principle described in UFC-3-340-02 (2008) is used to determine the adequacy of the designed elements for estimated blast loads with the consideration of the protection category required. Step by step explanation is given below and calculations are given in Annexes 2.1 to 2.3.

28

2.5.1 Design for flexure

Step 01:
Determine the loading arrangement of the elements based on BS 6399-1:1996 or Eurocodes (EN 1991 series).

Step 02:
Estimate reinforced concrete member sections using BS 8110-1:1997. Determine the ratio of steel to concrete ($\rho_l = A_s/A_c$).

Step 03:
Determine appropriate weights of explosives and stand off distances, and thereby estimate the blast load parameters. Values given in Table 2.2 and Annex 1.1 (Kingery and Bulmash solutions) are used in this study.

Step 04:
Determine the protection category. Depending on the required protection category, select the limiting value for θ and the design regime (quasi static dynamic or implusive).

(a) Design for protection category 1 (quasi static dynamic)

For protection category 1, θ is in the range 0~2°, section shall be type 1 where concrete is effective on both tensile and compressive surfaces and can be a single or double reinforced section.

Step 05 (a):

Define resistance deflection function in terms of ultimate resistance of the element R_m ($R_m = f(M_{Rd}, L)$). M_{Rd} is the design bending moment and L is the length of the span.

For cantilever slabs, from Annex 1.3, $R_m = 2M_p/L$ where M_p is the ultimate hogging moment capacity at supports.
For type 1 section, $M_{Rd} = A_s f_{yd,dyn}(d-0.5\lambda x)/b$
The depth from compression face to nutral axis, $x = A_s f_{yd,dyn}/(b\eta\lambda f_{cd,dyn})$
For $f_{ck} < 50$ MPa, $\lambda = 0.8$ and $\eta = 1.0$.
Substituting appropriate values for $f_{yd,dyn}$ and $f_{cd,dyn}$ from Table 2.4,
$M_{Rd} = 1.2\ A_s f_{yk}(d-0.4x)/b$, $x = A_s f_{yk}/(0.59\ b\ f_{ck})$
Then substituting values for A_s, f_{yk}, b, f_{ck}, value of x can be obtained and hence M_{Rd} can be found.
Since $M_{Rd} = M_p$, substituting values, R_m can be calculated.

Step 06 (a):
Calculate the unit resistance, $r_u = R_m/L_x$, and then calculate r_u/P_r.

Step 07 (a):
Calculate the maximum allowable deflection $x_{m\ all}$
Using an appropriate value for θ ($< 2°$), $\theta < 1°$ when there is no reinforcement as (blast) links, calculate the total deflection $x_{m\ all}$. (for a cantilever, $x_{m\ all} = L_x$ Tan θ).

Step 08 (a):
Estimate the deflection at elastic limit, $x_e = R_m/k_e$ as follows.
Using appropriate values for E_s, E_c and $\rho_l = A_s/A_c$, obtain the cofficient for second moment of area for cracked section (F) from Annex 1.4 as appropriate.
Calculate the second moment of area for cracked section $I_c = Fbd^3$
Then calculate the equivelent elastic stiffness for cracked section, k_e. Use appropriate equation for k_e from Annex 1.3. (For a cantilever, $k_e = 8E_cI_c/L^3$).
Then calculate x_e substituting values for R_m and k_e.

Step 09 (a):
Calculate the natural period of the element $T = 2\pi\sqrt{(K_{LM}M/k_e)}$ and hence t_d/T as follows.
From Annex 1.3, find the load mass factor K_{LM} ($K_{LM} = 0.65$ for cantilever with relevent loading arrangement).
Calculate the mass of the element M ($M = \rho_c hbL$ for rectangular element where ρ_c is the density of concrete).
Then calculate T substituting the values for K_{LM}, M and k_e and calculate t_d/T.

Step 10 (a):
Calculate the maximum deflection x_m and check with the maximum allowable deflection at the limit of θ (Step 07 (a)) as follows.
Obtain the maximum deflection x_m of elasto-plastic SDOF system for trianguler blast loads from Annex 1.5 using values for r_u/P_r (step 06 (a)) and t_d/T (step 09 (a)). When $x_m < x_{m\ all}$, the design is satisfactory.

Step 11 (a):
Check whether the response of the element is within quasi static – dynamic regime as follows.
Obtain the value for t_m/t_d for elasto-plastic SDOF system for triangular blast load from Annex 1.6 using values for r_u/P_r (step 06 (a)) and t_d/T (step 09 (a)). When $t_m/t_d < 3.0$, the design is satisfactory.

(b) Design for protection category 2 (impulsive)

For protection category 2, θ is more than $2°$, section is type 2 where concrete is cracked at tensile face and crushed at compressive face. Therefor, type 2 section must be a double reinforced section, i.e., reinforcement should be provided for the tensile side as well as for the compressive side of the section.

Step 05 (b):

Define resistance-deflection function in terms of ultimate resistance of the element R_m ($R_m = f(M_{Rd}, L)$). M_{Rd} is the design bending moment and L is the length of the span.

For cantilever slabs, from Annex 1.3, $R_m = 2M_p/L$ where M_p is the ultimate hogging moment capacity at support ($= M_{Rd}$).

For type 2 section, $M_{Rd} = A_s f_{yd,dyn}\, z/b$ where z is the lever arm (distance between tensile and compressive reinforcement).

As concrete is effective only within the depth z, steel to concrete ratio $\rho_1 = A_s/bz$

Therefore, $M_{Rd} = \rho_1 f_{yd,dyn}\, z^2$

Then, R_m can be calculated as, $R_m = (2/L)\,(\rho_1 f_{yd,dyn}\, z^2)$

Substituting appropriate value for $f_{yd,dyn}$ from Table 2.4, ($f_{yd,dyn} = 1.20\, f_{yk}$),

$R_m = (2.4/L)\,(\rho_1 f_{yk}\, z^2)$

Then substituting values for L, ρ_1 and f_{yk}, calculate R_m as a function of z,

$R_m = f(z)$.

Step 06 (b):

Estimate the allowable deflection $x_{m\,all}$ using an appropriate value for θ ($2° < \theta < 5°$), depending on (blast) links etc., (for cantilever, $x_m\,(= x_{m\,all}) = L_x\,Tan\,\theta$).

Step 07 (b):

Estimate the deflection at elastic limit $x_e = R_m/k_e$ as follows.

Using appropriate values for E_s, E_c and $\rho_1 = A_s/A_c$ (where $A_c = bz$), obtain the cofficient for second moment of area for cracked section (F) from Annex 1.4 as appropriate.

Calculate the second moment of area for cracked section $I_c = Fbz^3$ as a function of z.

Then calculate the equivelent elastic stiffness for cracked section, k_e as a function of z.

Use the appropriate equation for k_e from Annex 1.3. (For a cantilever, $k_e = 8E_c I_c/L^3$).

Then calculate x_e substituting values for R_m and k_e as a function of z, $x_e = f(z)$.

Step 08 (b):
Determine the load mass factor K_{LM} and mass of the element M as follows.
From Annex 1.3, find the load mass factor K_{LM} ($K_{LM} = 0.66$ for cantilever with relevent loading arrangement).
Calculate the mass of the element M ($M = \rho_c zbL$ for rectangular element where ρ_c is the density of concrete).

Step 09 (b):
Substituting values calculated above in the basic impulsive equation, find solutions for z.
$$I_r^2 A^2/(2K_{LM}M) = R_m x_e/2 + R_m/(x_m\text{-}x_e)$$
I_r is reflected impluse of blast, and A is the area loaded by blast pressure (Lb).

Step 10 (b):
Determine the modified requirement of steel (to resist blast loading).
Steel area, $A_s = \rho_1 z b$
Modify the bar diameter, spacing of bars etc., and the thickness of the element D if required.

Step 11 (b).
Check whether the response of the element is within implusive regime as follows.
Calculate the unit resistance, $r_u = R_m/L_x$ substituting z in the equation $R_m = f(z)$.
Then calculate the time at maximum deflection $t_m = I_r/r_u$.
Obtain the value for t_m/t_d.
When $t_m/t_d > 3.0$, the design is satisfactory.

Step 12:
For both quasi static – dynamic and implusive regimes,
Repeat the design with modified steel ratio ($\rho_1 = A_s/A_c$) if the design is not satisfactory.

2.5.2 Design for shear

The shear reinforcement design can be done according to BS 8110-1:1997 with the dynamic increase factors and dynamic design strengths mentioned in Tables 2.3 and 2.4.

CHAPTER 3

METHODOLOGY, CALCULATIONS AND RESULTS

3.1 Description

The method of design and steps followed in the study are explained below. Research was carried out for the following 3 categories of design.

   i.   Conventional design (designing using conventional codes without applying blast loading)
   ii. Blast resistant design for impulsive regime
   iii. Blast resistant design for quasi-static and dynamic regime

The loading arrangement for conventional designs was selected according to BS 6399-1:1996. For blast resistant designs, it was assumed that the cantilevered slab is always located above the point of explosion and therefore the blast loading on the cantilever slab is always from the underside of the slab (Figure 3.1). Reduction in shear forces and bending moments of blast loading due to downward dead and imposed loading was not considered in these designs.

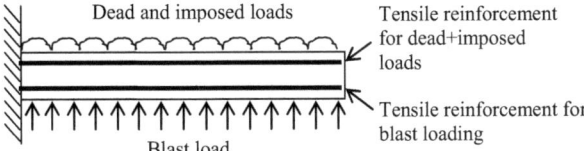

Figure 3.1: Loading diagram for a cantilever slab

3.2 Conventional design

Cantilevered slabs of span 1,000 mm, 1,500 mm, 2,000 mm and 3,000 mm were designed according to BS 6399-1:1996 and BS 8110-1:1997. Imposed load was considered up to its maximum (5 kN/m²) with dead load according to the designed thickness of slabs. The characteristic strength of concrete of 35 N/mm² and that of steel of 460 N/mm² were used in the calculations. The results of the designs are shown in Table 3.1 and the spread sheet used for detailed calculations is attached in Annex 2.1.

Table 3.1: Reinforcement detail for conventional designs

| Span (mm) | Conventional Design | | | |
	d_{eff} (mm)	A_s (mm²), ($A'_s = 0$ mm²)*	A_s/bd	Shear links (mm²/m²)
1,000	110	201	0.0018	0.00
1,500	165	342	0.0021	0.00
2,000	220	511	0.0023	0.00
3,000	330	931	0.0028	0.00

*: A_s' is the compressive reinforcement area

Points to be highlighted from the results are; that the effective depth can be maintained within practical limits (110 mm to 330 mm) with steel / concrete ratio of 0.0018 to 0.0028 and that shear design is satisfied without additional shear links.

3.3 Blast resistant design for impulsive regime

Impulsive regime exists in protection category 2 where only structural elements are safe from collapse. Therefore, if any conventional design is within the impulsive boundries, it can be determined that at least the structure is safe.

In impulsive designs, support rotation θ is allowed more than 2° (up to 4°). This higher support rotations and deflections cause cracking of concrete on the tensile face of the element and crushing on the compressive face. Therefore, impulsive designs require reinforcements on both compressive and tensile faces. It is recommended that at least minimum reinforcement be provided on the compressive face of the element. Especially for cantilever designs, the dead and imposed loads act on the top side and most probably the blast load acts on the underside. Therefore steel is required at both sides (Figure 3.1).

Designs were done for spans of 1,000 mm, 1,500 mm, 2,000 mm and 3,000 mm and steel to concrete ratios of 0.0005, 0.00075, 0.001, 0.0015, 0.002, 0.003, 0.004, 0.005, 0.006, 0.007, 0.008, 0.009, 0.01, 0.02 in each span and for each Z value for the range from 0.4310 to 40.9405 m/kg$^{1/3}$. The calculations were done using a spread sheet. A specimen calculation for 1,000 mm cantilever slab design is given in Annex 2.2.

For clarity, the results of design calculations, i.e., A_s, A'_s, $(A_s+A'_s)/(bd)$, d_{eff}, are given in two tables: Table 3.2 shows selected results for each span at $Z = 2.155$ m/kg$^{1/3}$ and Table 3.3 shows the complete results of the impulsive regime calculations. Results of Table 3.1 of conventional design can be compared with the results of Table 3.2 of implusive design in order to understand the difference between the two designs.

Table 3.2: Reinforcement detail for impulsive design at $Z = 2.155$ m/kg$^{1/3}$

Span (mm)	Impulsive design			
	d_{eff} (mm)	$A_s+A'_s$ (mm^2)	$(A_s+A'_s)/(bd)$	Shear links (mm^2/m^2)
1,000	248	425	0.002	0.00
1,500	229	777	0.004	0.00
2,000	245	858	0.004	0.00
3,000	265	1150	0.005	0.00

Table 3.3: Impulsive design, value of Z for $0.431 < Z < 40.94$ m/kg$^{1/3}$

(a) Cantilever slabs with span 1,000 mm

Span, d_{eff} (mm)	(As)/bd*														
	0.0005	0.00075	0.0010	0.0015	0.002	0.003	0.004	0.005	0.006	0.007	0.008	0.009	0.01	0.015	0.02
1000, 70	40.94	Fail	Fail	Fail	Fail	Fail	Fail	Fail	Fail	Fail	Fail	Fail	Fail	Fail	Fail
1000, 75	31.24	Fail	Fail	Fail	Fail	Fail	Fail	Fail	Fail	Fail	Fail	Fail	Fail	Fail	Fail
1000, 100	15.08	Fail	Fail	Fail	Fail	Fail	Fail	Fail	Fail	Fail	Fail	Fail	Fail	Fail	Fail
1000, 150	6.464	Fail	Fail	Fail	Fail	Fail	Fail	Fail	Fail	Fail	Fail	Fail	Fail	Fail	Fail
1000, 200	3.879	Fail	Fail	Fail	Fail	Fail	Fail	Fail	Fail	Fail	Fail	Fail	Fail	Fail	Fail
1000, 250	2.801	2.370	2.155	Fail	Fail	Fail	Fail	Fail	Fail	Fail	Fail	Fail	Fail	Fail	Fail
1000, 300	2.155	1.869	1.652	Fail	Fail	Fail	Fail	Fail	Fail	Fail	Fail	Fail	Fail	0.613	0.568
1000, 350	1.796	1.508	1.347	1.025	0.977	0.882	0.818	0.754	0.704	0.646	0.632	0.616	0.600	0.530	0.470
1000, 400	1.465	1.257	1.149	0.985	0.862	0.776	0.694	0.634	0.608	0.579	0.560	0.539	0.523	0.448	0.372

Span, d_{eff} (mm)	0.0005	0.00075	0.0010	0.0015	0.002	0.003	0.004	0.005	0.006	0.007	0.008	0.009	0.01	0.015	0.02
1000, 450	1.262	1.077	1.131	0.845	0.772	0.668	0.612	0.575	0.545	0.512	0.488	0.462	0.446	0.365	0.274
1000, 500	1.108	0.970	0.862	0.762	0.682	0.605	0.556	0.515	0.431	0.445	0.417	0.385	0.370	0.282	0.176

(b) Cantilever Slabs with span 1,500 mm

Span, d_{eff} (mm)	(As)/bd*														
	0.0005	0.00075	0.0010	0.0015	0.002	0.003	0.004	0.005	0.006	0.007	0.008	0.009	0.01	0.015	0.02
1500, 65	40.94	40.94	40.94	40.94	Fail	Fail	Fail	Fail	Fail	Fail	Fail	Fail	Fail	Fail	Fail
1500, 75	39.86	33.99	29.08	22.62	Fail	Fail	Fail	Fail	Fail	Fail	Fail	Fail	Fail	Fail	Fail
1500, 100	17.23	15.08	12.92	10.77	10.77	Fail	Fail	Fail	Fail	Fail	Fail	Fail	Fail	Fail	Fail
1500, 150	8.080	6.464	5.387	4.741	4.094	Fail	Fail	Fail	Fail	Fail	Fail	Fail	Fail	Fail	Fail
1500, 200	4.741	4.094	3.448	3.017	2.586	2.480	1.939	1.836	Fail	Fail	Fail	Fail	Fail	Fail	Fail
1500, 250	3.448	2.801	2.586	2.155	1.939	1.616	1.437	1.293	Fail	Fail	Fail	Fail	Fail	Fail	Fail
1500, 300	2.586	2.155	1.939	1.652	1.508	1.250	1.131	1.042	0.991	0.948	0.862	0.835	0.808	0.708	0.646
1500, 350	2.083	1.796	1.580	1.347	1.207	1.047	0.934	0.862	0.814	0.916	0.742	0.700	0.673	0.600	0.557
1500, 400	1.724	1.465	1.336	1.149	1.024	0.893	0.803	0.742	0.694	0.646	0.634	0.604	0.593	0.523	0.467
1500, 450	1.508	1.262	1.149	0.997	0.889	0.784	0.705	0.636	0.612	0.587	0.570	0.545	0.525	0.446	0.377
1500, 500	1.293	1.108	1.012	0.862	0.796	0.686	0.626	0.585	0.556	0.527	0.507	0.482	0.458	0.370	0.287

(c) Cantilever slabs with span 2,000 mm

Span, d_{eff} (mm)	(As)/bd*														
	0.0005	0.00075	0.0010	0.0015	0.002	0.003	0.004	0.005	0.006	0.007	0.008	0.009	0.01	0.015	0.02
2000, 60	40.94	40.94	40.94	40.94	40.94	40.94	Fail	Fail	Fail	Fail	Fail	Fail	Fail	Fail	Fail

Span, d_{eff} (mm)	0.0005	0.00075	0.0010	0.0015	0.002	0.003	0.004	0.005	0.006	0.007	0.008	0.009	0.01	0.015	0.02
2000, 75	40.94	39.86	33.39	29.08	24.78	20.47	18.31	Fail	Fail	Fail	Fail	Fail	Fail	Fail	Fail
2000, 100	21.54	17.23	15.08	12.92	10.77	8.619	7.542	7.542	Fail	Fail	Fail	Fail	Fail	Fail	Fail
2000, 150	9.696	7.542	6.464	5.387	4.741	4.094	3.663	3.232	3.017	2.801	2.586	2.586	2.370	Fail	Fail
2000, 200	5.387	4.525	4.094	3.448	3.017	2.586	2.263	2.047	1.939	1.832	1.724	1.616	1.616	Fail	Fail
2000, 250	3.879	3.232	2.801	2.370	2.155	1.832	1.616	1.508	1.401	1.293	1.239	1.221	1.149	0.991	0.916
2000, 300	3.017	2.478	2.155	1.868	1.652	1.437	1.293	1.185	1.077	1.034	0.991	0.948	0.905	0.800	0.718
2000, 350	2.370	2.011	1.795	1.508	1.347	1.164	1.047	0.970	0.898	0.844	0.814	0.790	0.754	0.646	0.600
2000, 400	1.939	1.670	1.508	1.262	1.149	0.985	0.893	0.823	0.776	0.733	0.694	0.670	0.634	0.575	0.523
2000, 450	1.681	1.437	1.293	1.108	0.997	0.845	0.784	0.725	0.668	0.636	0.614	0.599	0.575	0.503	0.446
2000, 500	1.472	1.262	1.139	0.982	0.862	0.762	0.686	0.637	0.605	0.582	0.556	0.539	0.515	0.431	0.370

(d) Cantilever slabs with span 3,000 mm

Span, d_{eff} (mm)	(As)/bd*														
	0.0005	0.00075	0.0010	0.0015	0.002	0.003	0.004	0.005	0.006	0.007	0.008	0.009	0.01	0.015	0.02
3000, 65	40.94	40.94	40.94	40.94	40.94	40.94	40.94	Fail	Fail	Fail	Fail	Fail	Fail	Fail	Fail
3000, 75	40.94	40.94	40.94	36.63	32.32	26.93	23.70	22.62	21.54	19.39	19.39	19.39	Fail	Fail	Fail
3000, 100	28.01	23.16	20.47	17.23	15.08	12.12	10.77	9.816	8.619	8.619	7.973	7.542	7.542	6.195	Fail
3000, 150	11.69	9.760	8.423	7.003	6.177	5.171	4.525	4.094	3.792	3.577	3.361	3.196	3.017	2.585	2.37
3000, 200	6.883	5.710	4.999	4.166	3.663	3.088	2.737	2.488	2.316	2.155	2.072	1.939	1.896	1.622	1.457
3000, 250	4.741	3.933	3.448	2.909	2.586	2.155	1.939	1.778	1.661	1.565	1.508	1.421	1.365	1.175	1.058

3000, 300	3.546	2.974	2.640	2.223	1.939	1.681	1.508	1.382	1.293	1.221	1.161	1.105	1.062	0.924	0.834
3000, 350	2.464	2.370	1.857	1.791	1.603	1.368	1.229	1.128	1.055	1.003	0.955	0.913	0.873	0.770	0.694
3000, 400	2.085	1.975	1.579	1.508	1.344	1.154	1.038	0.959	0.894	0.846	0.813	0.784	0.756	0.644	0.602
3000, 450	1.801	1.688	1.372	1.293	1.162	1.000	0.824	0.830	0.786	0.746	0.709	0.675	0.646	0.583	0.534
3000, 500	1.587	1.474	1.216	1.132	1.021	0.872	0.742	0.741	0.692	0.646	0.627	0.608	0.592	0.522	0.466

*: $A_s/(bd)$ mentioned in the table is only A_s (without A_s'). However for the design, the actual reinforcement requirement is $A_s + A_s'$ for impulsive regieme.

"Fail" indicates that the design does not satisfy the quasi-static and dynamic design criteria.

In this study, the limits of d_{eff} and $A_s/(bd)$ used for the calculations are:

- Limits of d_{eff}; Minimum: 65 mm to Maximum: 400 mm
- Limits of $A_s/(bd)$; Minimum: 0.0005 to Maximum: 0.020 (The details available on $A_s/(bd)$ for blast designs are limitted to 0.02).

The results shown in Table 3.3 describe the relationship between the $A_s/(bd)$ ratio of cantilevered reiforced concrete sections and Z with respect to varying d_{eff} for impulsive regime. These design envelopes of the impulsive regime are graphically presented in Figures 3.2 – 3.5.

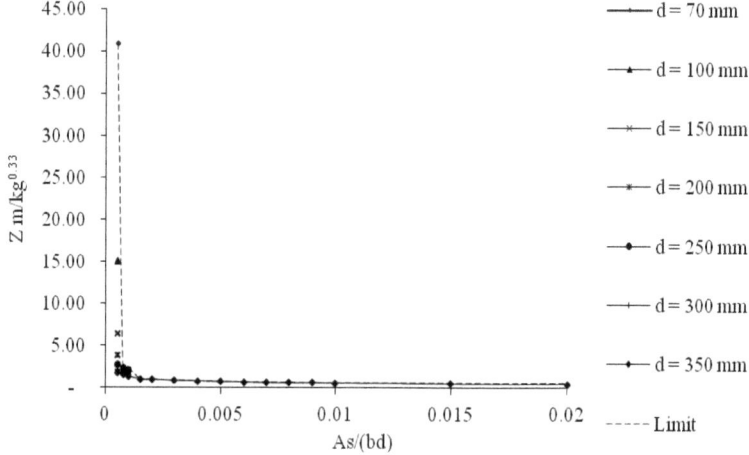

Figure 3.2: Envelope for cantilevers of span 1,000 mm in impulsive regime

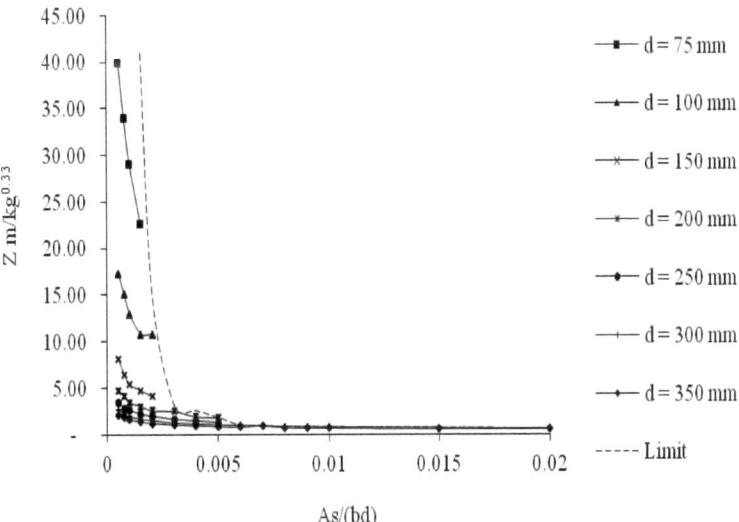

Figure 3.3: Envelope for cantilevers of span 1,500 mm in impulsive regime

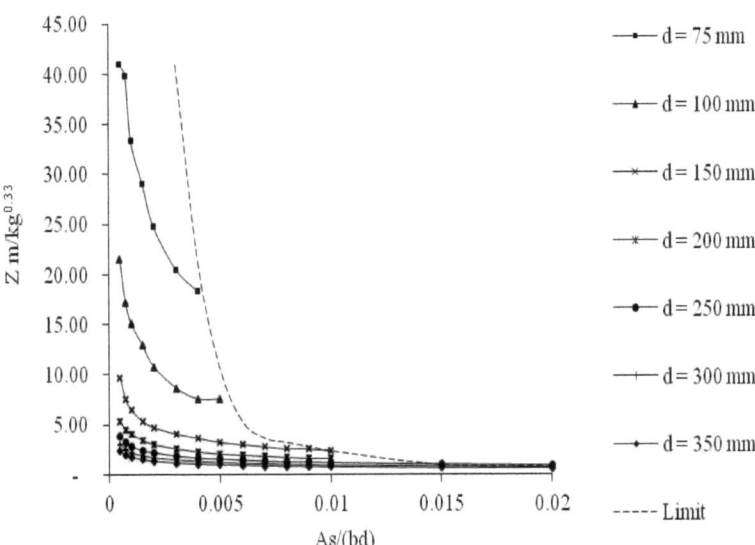

Figure 3.4: Envelope for cantilevers of span 2,000 mm in impulsive regime

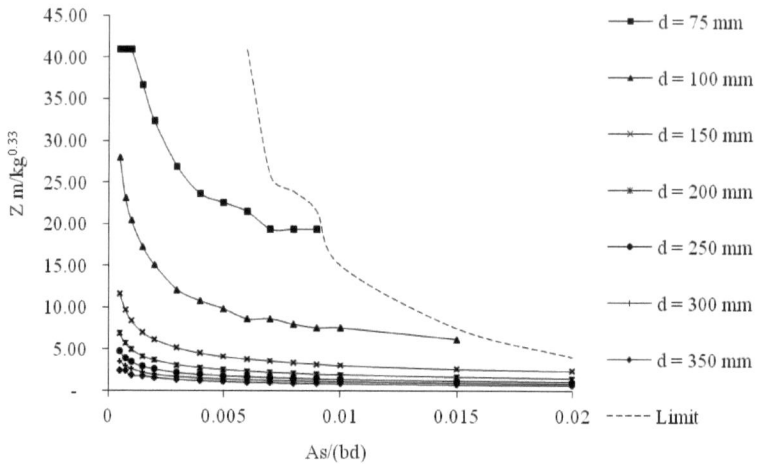

Figure 3.5: Envelope for cantilevers of span 3,000 mm in impulsive regime

In Figures 3.2 – 3.5, "d" means the effective depth. "Limit" means that the design beyond this limit does not satisfy impulsive design criteria.

3.4 Blast resistant design for quasi static and dynamic regime

Quasi static and dynamic regime is the regime for protection category 1 designs that expect the protection of occupants, furniture or equipment together with structural elements. Therefore, if any conventional design is within the quasi-static and dynamic regime, it can be determined that the structure and occupants are safe for the specific range of blast magnitudes.

In quasi static and dynamic designs, the support rotation θ must be less than 2°. Therefore, crushing of concrete is not expected on the compressive face of elements though cracking is allowed on the tensile face. Accordingly, in these designs, reinforcement is theoritically required only on the tensile face. However on cantilevers, the dead and imposed loads act on the top side of the cantilevers and the blast load is most likely to act on the underside. Therefore, steel is required for both top and bottom faces.

In this study, designs were carried out for spans 1,000 mm, 1,500 mm, 2,000 mm and 3,000 mm and steel to concrete ratios of 0.0005, 0.00075, 0.001, 0.0015, 0.002, 0.003, 0.004, 0.005, 0.006, 0.007, 0.008, 0.009, 0.01, 0.02 in each span and for each Z value for the range from 0.4310 to 40.9405 m/kg$^{1/3}$.

The calculations were done by using a spread sheet; a specimen calculation for 1,000 mm spanned cantilever slab is given in Annex 2.3.

Unlike in the impulsive regime designs, in quasi static and dynamic regime, the calculations were done for different effective depths as per the design methodology. The results of the design calculations, i.e., A_s, A'_s, $(A_s+A'_s)/(bd)$, shear links and d_{eff} are given in two tables. Table 3.4 shows selected results for each span at $Z = 2.155$ m/kg$^{1/3}$ and Table 3.5 shows the complete results of the quasi static and dynamic regime calculations. The differences between convensional designs, impulsive designs and quasi static and dynamic designs become clear when the results given in Tables 3.1, 3.2 and 3.4 are compared.

Table 3.4: Reinforcement detail for quasi static and dynamic design at $Z = 2.155$ m/kg$^{1/3}$

Span (mm)	Quasi static, dynamic regime			
	d_{eff} (mm)	$A_s+A'_s$ (mm^2)	$(A_s+A'_s)/(bd)$	Shear links (mm^2/m^2)
1,000	215	3440	0.016	3,934
1,500	255	4590	0.018	4,647
2,000	280	5600	0.020	3,203
3,000	350	10500	0.030	2,878

Table 3.5: Quasi static and dynamic design, value of Z for $0.431 < Z < 40.94$ m/kg$^{1/3}$

Span, d_{eff} (mm)	As/bd*													
	0.0005	0.0008	0.0010	0.0020	0.0030	0.0040	0.0050	0.0060	0.0070	0.0080	0.0090	0.0100	0.0150	0.0200
1000, 100	13.41	11.49	9.583	6.542	6.270	5.441	4.995	4.740	4.518	4.295	4.156	4.016	3.551	3.248
1000, 150	7.171	6.469	5.767	4.446	3.998	3.609	3.315	3.202	3.061	2.919	2.838	2.756	2.447	2.305
1000, 200	5.146	4.753	4.360	3.432	3.040	2.801	2.563	2.482	2.399	2.316	2.245	2.173	2.013	1.913
1000, 250	4.358	3.932	3.506	2.846	2.533	2.330	2.205	2.122	2.074	2.026	1.973	1.920	1.728	1.683
1000, 300	3.842	3.454	3.065	2.462	2.214	2.066	1.973	1.906	1.866	1.826	1.767	1.708	1.608	1.498

1500, 100	22.27	17.96	13.65	9.604	7.860	6.966	6.069	5.911	5.541	5.171	5.004	4.837	4.439	4.073
1500, 150	12.43	9.953	7.471	5.607	4.819	4.308	3.988	3.861	3.769	3.676	3.570	3.464	3.079	2.864
1500, 200	8.023	6.597	5.171	4.017	3.570	3.302	3.125	2.910	2.834	2.757	2.662	2.566	2.334	2.241
1500, 250	6.269	5.202	4.134	3.315	3.026	2.775	2.609	2.472	2.389	2.305	2.267	2.228	2.077	1.927
1500, 300	4.956	4.256	3.556	2.785	2.522	2.370	2.266	2.180	2.140	2.100	2.055	2.010	1.847	1.741
2000, 100	Fail	Fail	32.32	14.37	10.42	8.352	7.201	7.048	6.650	6.251	6.125	5.999	5.187	4.765
2000, 150	30.16	21.54	12.93	7.542	5.748	5.372	4.809	4.601	4.386	4.170	4.075	3.980	3.429	3.268
2000, 200	14.01	11.14	8.264	5.171	4.310	3.911	3.668	3.563	3.376	3.189	3.126	3.062	2.801	2.586
2000, 250	9.798	7.809	5.820	4.094	3.448	3.169	2.801	2.738	2.632	2.526	2.435	2.344	2.179	2.111
2000, 300	7.237	5.989	4.740	3.520	2.974	2.705	2.457	2.372	2.289	2.205	2.158	2.111	1.990	1.892
3000, 100	Fail	Fail	Fail	Fail	Fail	38.79	30.12	28.01	23.70	19.39	17.24	15.08	8.619	6.831
3000, 150	Fail	Fail	Fail	38.77	19.39	12.93	8.619	7.542	7.003	6.464	5.945	5.425	4.525	3.900
3000, 200	Fail	Fail	36.63	12.22	8.933	6.538	5.659	5.021	4.673	4.324	4.124	3.923	3.319	2.987
3000, 250	Fail	30.17	16.91	7.908	6.109	4.849	4.239	3.952	3.706	3.459	3.310	3.161	2.740	2.532
3000, 300	30.17	15.08	10.77	5.928	4.598	3.966	3.521	3.362	3.147	2.931	2.836	2.740	2.414	2.198

*: $A_s/(bd)$ mentioned in the table is only A_s (without A_s'). However for the design, the actual reinforcement requirement may be $A_s + A_s'$ depending on loads etc. The details available on steel / concrete ratio for blast designs are limitted to 0.02.

"Fail" indicates that the design does not satisfy quasi-static and dynamic design criteria.

In this study, the limits of d_{eff} and $A_s/(bd)$ used for the design calculations are:

- o Limits of d_{eff}; from 100 mm to 300 mm
- o Limits $A_s/(bd)$; from 0.0005 to 0.020

The results shown in Table 3.5 describe the relationship between $A_s/(bd)$ ratio of cantilevered reiforced concrete sections and Z with respect to varying d_{eff} for quasi static and dymanic regime. These design envilepes of the quasi static and dynamic regime are graphically presented in Figures 3.6 – 3.9.

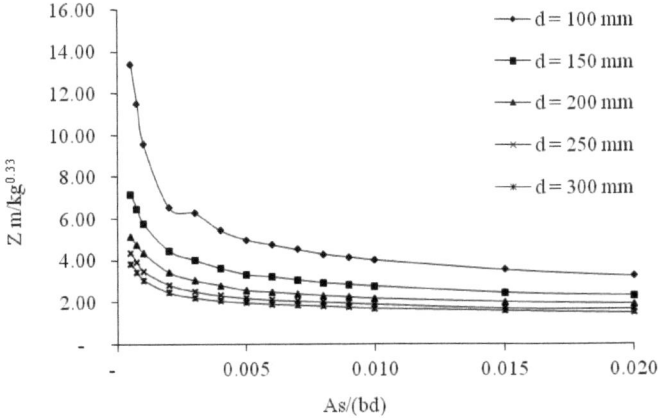

Figure 3.6: Envelope for cantilevers of span 1,000 mm in quasi static dynamic regime

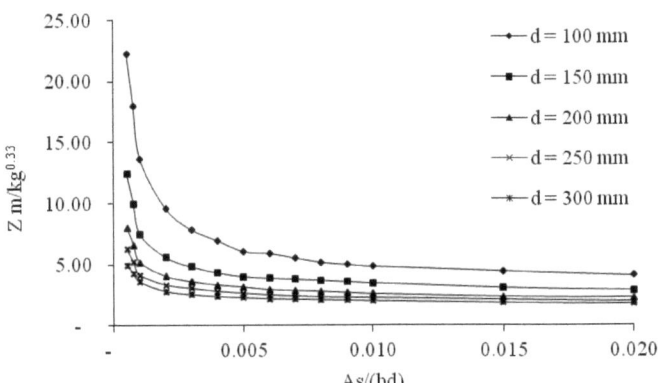

Figure 3.7: Envelope for cantilevers of span 1,500 mm in quasi static dynamic regime

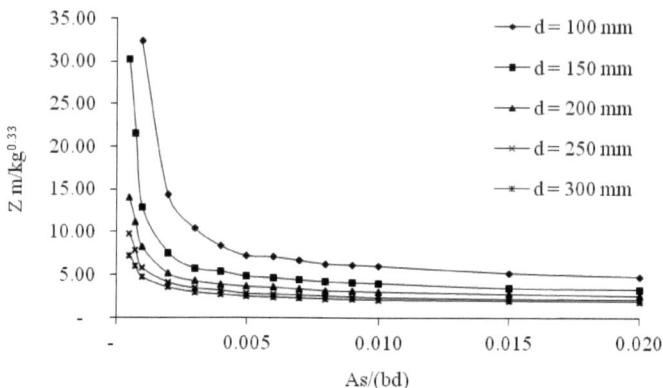

Figure 3.8: Envelope for cantilevers of span 2,000 mm in
quasi static dynamic regime

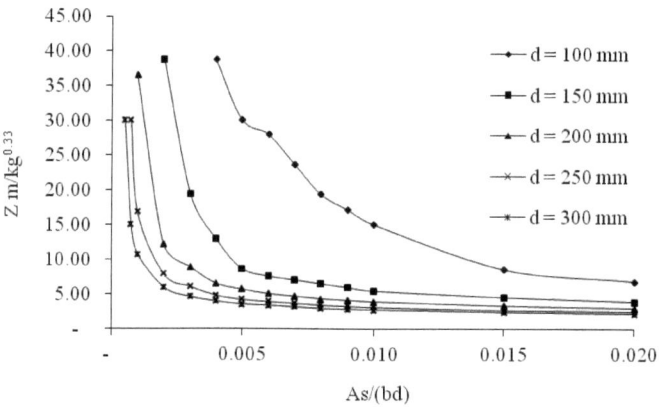

Figure 3.9: Envelope for cantilevers of span 3,000 mm in
quasi static dynamic regime

"d" in Figures 3.6 – 3.9 denotes the effective depth of the reinforced concrete section.

3.5 Analysis of results

Z that described the magnitude of blast selected in Tables 3.2 and 3.4, $Z = 2.155$
$m/kg^{1/3}$ is equivalent to 100 kg of TNT exploded at a distance of 10 m from the
considered element. In general, design for Z, approximately around 2 $m/kg^{1/3}$, is
considered sufficient for most of the structure.

Comparing the results given in Tables 3.1, 3.2 and 3.4, it is clear that the conventional designs can be improved to better withstand a blast (impulsive regime) with the scaled distance of $Z = 2.155$ m/kg$^{1/3}$ by using slight modifications including reinforcement at both compression and tensile sides. However when the span of the element is lesser (i.e., at 1,000 mm etc), greater thickness is required to increase the weight of the element.

Complete results given in Table 3.5 show that the conventional designs are within the quasi static and dynamic regime but with less blast resistant ability. It was also observed that, in order to improve a conventional design to quasi static and dynamic design, many modifications are necessary such as additional shear links and higher $A_s/(bd)$ ratios.

3.6 Comparing results of conventional and impulsive designs

The graphical presentation of impulsive regime results; $A_s/(bd)$ vs Z for selected range of effective depths are given in Figures 3.10 to 3.13.

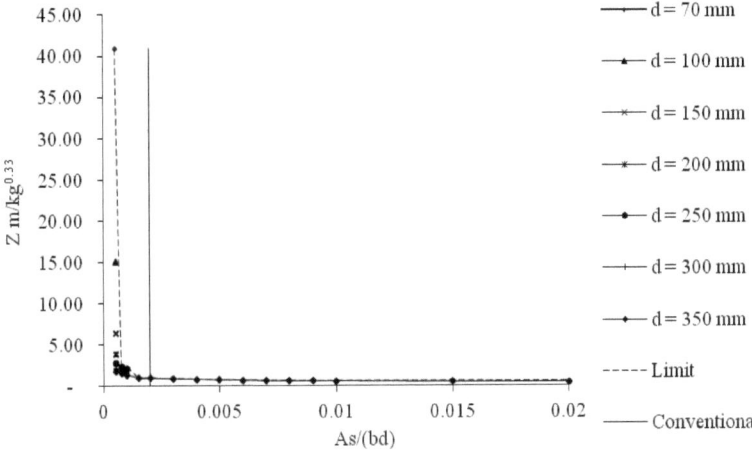

Figure 3.10: Position of conventional cantilever design in impulsive regime, span 1,000 mm

As shown in Figure 3.10, conventional design for a cantilever of span 1,000 mm is outside of impulsive regime ($d_{eff} = 110$ mm) and can be improved to be within the impulsive regime by increasing d_{eff} (> 350 mm). The effect of change of steel area for such elements is minor.

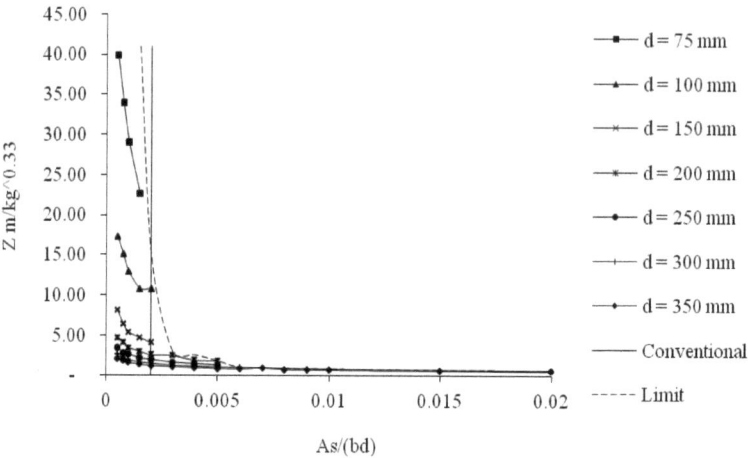

Figure 3.11: Position of conventional cantilever design in impulsive regime, span 1,500 mm

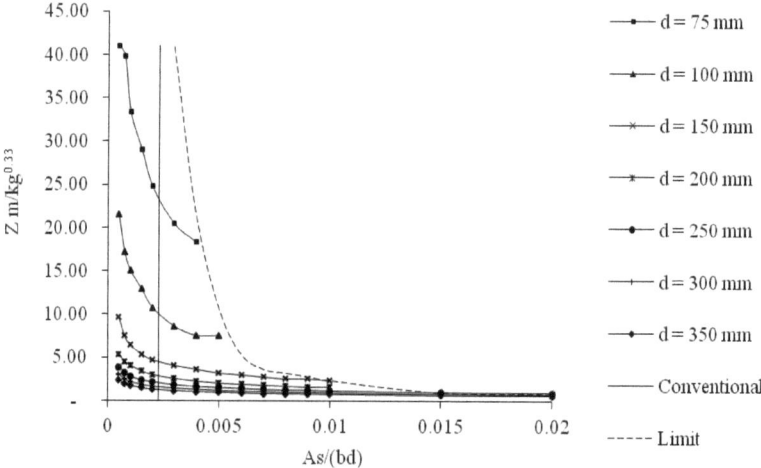

Figure 3.12: Position of conventional cantilever design in impulsive regime, span 2,000 mm

In the impulsive design envelope for span 1,500 mm cantilevers (Figure 3.11), conventional design stays within the limit of impulsive regime for d_{eff} = 165 mm with $A_s/(bd)$ = 0.002. However, reinforcements should be provided at both top and bottom faces ($A_s/(bd)$ = 0.004). Increasing of steel area will first improve the element's resistance to blasts (but it is not economical because Z gets reduced only slightly) and when $A_s/(bd)$ goes over the limit, it shifts the element away from the impulsive regime. The value of Z stays below 5 m/kg$^{0.33}$ (equivalent to a surface blast of 100 kg TNT at a distance of 24 m).

As in Figure 3.12, conventional design of span 2,000 mm cantilever stays within the impulsive regime for d_{eff} = 220 mm with $A_s/(bd)$ = 0.002. Reinforcement is required at both top and bottom faces ($A_s/(bd)$ = 0.004). Increasing of steel area will first improve the element's resistance to blasts (but it is not economical because Z gets reduced only slightly) and when $A_s/(bd)$ goes over the limit, it moves the element away from the impulsive regime. The value of Z stays below 3m/kg$^{1/3}$.

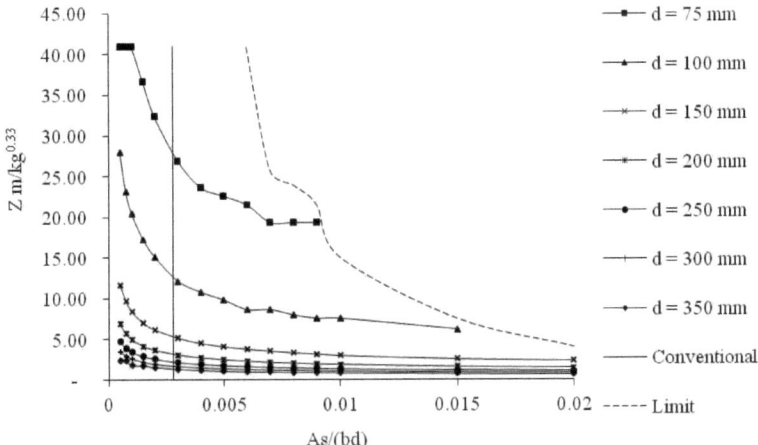

Figure 3.13: Position of conventional cantilever design in
impulsive regime, span 3,000 mm

Figure 3.13 shows that the conventional design for span 3,000 mm cantilever stays within the impulsive regime for d_{eff} = 330 mm with $A_s/(bd)$ = 0.003. Reinforcement is required at both top and bottom faces ($A_s/(bd)$ = 0.006). Increasing of steel area will improve the element's resistance to blasts (it is not economical because Z gets reduced only slightly). Value of Z stays below 2 m/kg$^{1/3}$ (equivalent to a surface blast of 100 kg TNT at a distance of 9 m).

The impulsive regime results show that, in order to improve the blast resistant properties of cantilever slabs, either d_{eff} has to be increased or both d_{eff} and $A_s/(bd)$ have to be increased. It is important to note that, the impulsive properties cannot be increased simply by increasing $A_s/(bd)$. For smaller effective depths (depending on the span), there is a maximum limit for $A_s/(bd)$. Once $A_s/(bd)$ goes beyond that limit, the element will no longer act in the impulsive regime. In general, improvements made to blast resistant properties by increasing the steel area in the impulsive regime have only a minor effect. Therefore, increasing steel in the impulsive regime is not economical.

3.7 Comparing results of conventional design and quasi static dynamic designs

The range of results, i.e., Z vs $A_s/(bd)$ with different d_{eff}, selected from Table 3.5 as mentioned below are plotted in Figure 3.14.
- o   Results for 1,000 mm span for d_{eff} 100 mm and 300 mm
- o   Results for 1,500 mm span for d_{eff} 100 mm and 300 mm
- o   Results for 2,000 mm span for d_{eff} 100 mm and 300 mm
- o   Results for 3,000 mm span for d_{eff} 100 mm and 300 mm

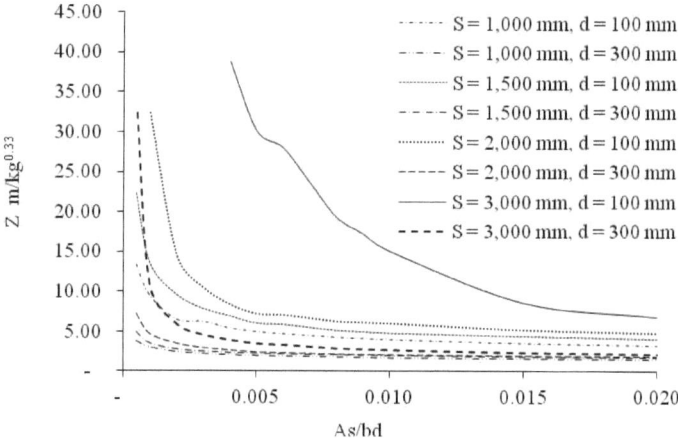

Figure 3.14: Quasi static and dynamic (QSD) range

"S" means span, "d" means effective depth

The quasi static and dynamic (QSD) graphs show the maximum and minimum limits of Z for each span corresponding to effective depths. Therefore any point between the graphs corresponding to that span of the cantilever is valid and safe in a blast with corresponding Z. The maximum $A_s/(bd)$ value is limited to 0.02

48

and the element may be reinforced for both tension and compression with shear links.

The position of a conventional cantilever slab design is compared with quasi-static and dynamic regime results graphically in Figures 3.15 to 3.18. They show the position of conventional design corresponding to each span.

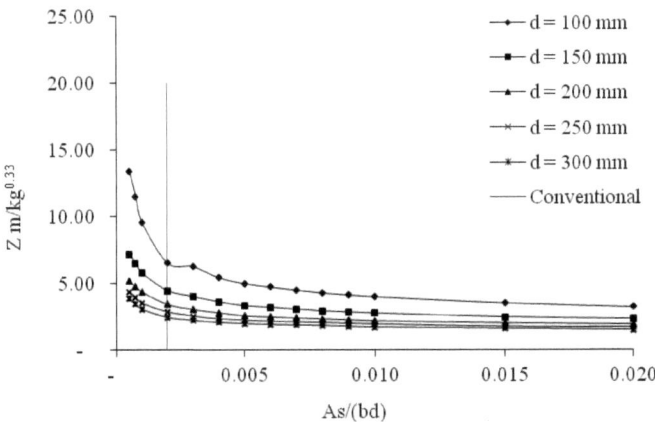

Figure 3.15: Position of conventional cantilever design in QSD regime, span 1,000 mm

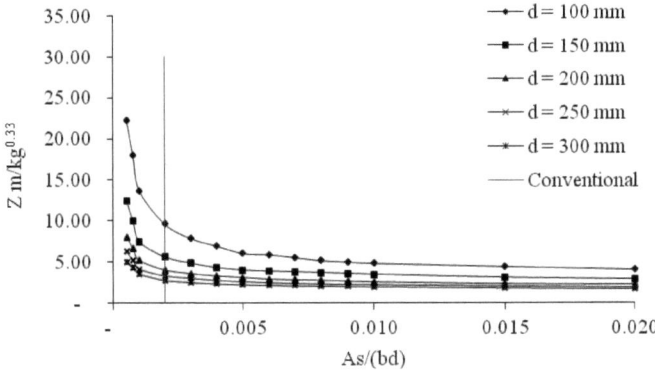

Figure 3.16: Position of conventional cantilever design in QSD regime, span 1,500 mm

Figure 3.15 shows that the conventional design of span 1,000 mm cantilevered slab stays within the quasi static and dynamic regime. The values of $A_s/(bd) =$

0.002 and d_{eff} =110 mm show Z around 7 m/kg$^{1/3}$ (equivalent to 100 kg TNT explosion on the surface at a distance of 33 m). Increasing of d_{eff} has a bigger influence for improving blast resistant properties than increasing of steel area.

The conventional design of 1,500 mm cantilever stays within the quasi static and dynamic regime (Figure 3.16). The values of $A_s/(bd)$ = 0.002 and d_{eff} =165 mm show Z less than 5 m/kg$^{1/3}$ (equivalent to a surface blast of 100 kg TNT at a distance of 24 m). As in the 1,000 mm span, increasing of d_{eff} has greater influence towards improving blast resistant properties than the increasing of steel area.

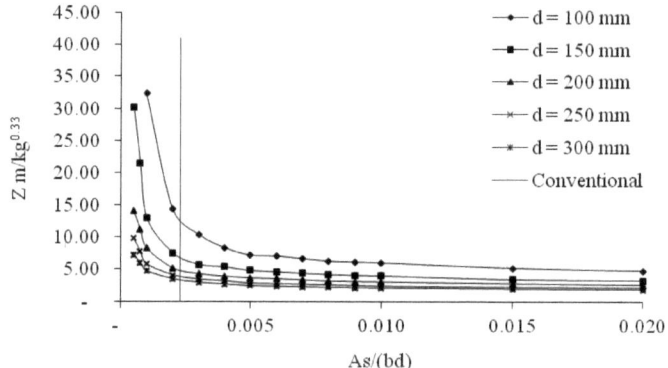

Figure 3.17: Position of the conventional cantilever design in
QSD regime, span 2,000 mm

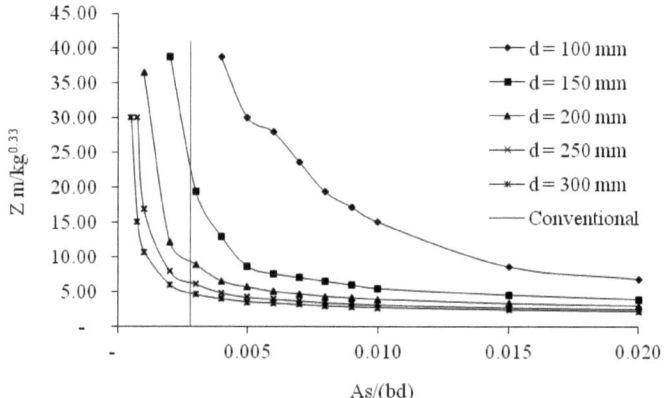

Figure 3.18: Position of conventional cantilever design in
QSD regime, span 3,000 mm

50

The conventional cantilever design for span 2,000 mm also stays within the quasi static and dynamic regime. The values of $A_s/(bd) = 0.002$ and $d_{eff} = 220$ mm show Z less than 5 m/kg$^{1/3}$. As in previous spans, the increase of d_{eff} has a bigger influence in improving blast resistant properties of the cantilever than the increase of steel area.

As observed in Figure 3.18, the conventional cantilever design with span 3,000 mm too stays within the quasi static and dynamic regime. The values of $A_s/(bd)$ = 0.003 and d_{eff} = 330 mm show Z less than 5 m/kg$^{1/3}$. Similar to previous spans, increasing of d_{eff} has a greater influence in improving blast resistant properties than increasing of steel area.

The quasi static and dynamic regime results show that, to improve the blast resistant properties of cantilever slabs, either d_{eff} has to be increased or both d_{eff} and $A_s/(bd)$ have to be increased. In the quasi static and dynamic regime too, it is important to note that, the blast resistant properties cannot be increased simply by increasing the steel area. In general, improvements done to blast resistant properties by increasing the steel area have only a minor effect. Therefore, increasing of steel is not economical. However, unlike in the impulsive regime, any conventional design stays within the quasi static and dynamic regime and the only concern is the magnitude of blast pressure the element can bear.

3.8 Analysis of impulsive and quasi static dynamic results

The following is a summary of the position of conventional designs in blasting environment for both impulsive and quasi static dynamic regimes.

i.   Comparing Figure 3.10 with Figure 3.15, for 1,000 mm spanned cantilever slabs, d_{eff} = 100 mm to 150 mm do not stay in the impulsive regime but stays in the quasi static and dynamic regime for range of Z from 7.0 m/kg$^{0.33}$ to 4.5 m/kg$^{0.33}$. However by increasing the effective depth (i.e. d_{eff} = 350 mm), these cantilevers can be pushed in to impulsive regime while the quasi static and dynamic properties too can be improved to withstand a blast with $Z = 2.2$ m/kg$^{0.33}$.

ii. Cantilever slabs with the span 1,500 mm, d_{eff} = 100 mm to 200 mm stay in the impulsive regime for the range of Z from 11.5 m/kg$^{0.33}$ to 2.5 m/kg$^{0.33}$. However, reinforcement is required at both tensile and compressive faces (Figures 3.11 and 3.16). These cantilevers stay in the quasi static and dynamic regime too for the range of Z from 9.5 m/kg$^{0.33}$ to 4.0 m/kg$^{0.33}$.

iii. Comparing Figure 3.12 with Figure 3.17, 2,000 mm spanned cantilever slabs with d_{eff} = 150 mm to 250 mm stay in the impulsive regime for the range of Z from 4.5 m/kg$^{0.33}$ to 2.0 m/kg$^{0.33}$. Reinforcement is required at both tensile and compressive faces. These cantilevers stay in the quasi static and dynamic regime too for the range of Z from 7.0 m/kg$^{0.33}$ to 3.5 m/kg$^{0.33}$.

iv. As Figures 3.13 and 3.18 show, 3,000 mm spanned cantilever slabs (d_{eff} = 200 mm to 350 mm) stay in the impulsive regime for the range of Z from 3.5 m/kg$^{0.33}$ to 2.0 m/kg$^{0.33}$. Here too, reinforcement is necessary at both tensile and compressive faces. d_{eff} = 200 mm to 300 mm stay in the quasi static and dynamic regime too for the range of Z from 9.5 m/kg$^{0.33}$ to 5.0 m/kg$^{0.33}$.

The magnitude of Z can be reduced by increasing the effective depth (slab thickness) and steel area but increase of effective depths gives better results. The main reason for this is that the higher the mass (weight) of the element, the higher the resistance to blast loading. Converting a conventional design into an impulsive design is more economical than converting it into a quasi static and dynamic design; this is mainly due to the requirement of shear links for quasi static and dynamic regime.

3.9 Shear reinforcement

Shear reinforcement is not usually used in conventional slab designs. However, in blast resistant designs, especially in the quasi static and dynamic designs, shear reinforcement plays a major role. A comparison of shear reinforcement between impulsive and quasi static and dynamic regimes for $Z = 2.155$ m/kg$^{1/3}$ is given in Table 3.6. Requirement of shear reinforcement for cantilevers of span 3,000 mm is given in Figure 3.19. It was observed that d_{eff} < 300 mm cannot be designed for $Z < 1.05$ m/kg$^{1/3}$ and that shear reinforcement is not required for $Z > 1.05$ m/kg$^{1/3}$.

Table 3.6: Shear reinforcement for $Z = 2.155$ m/kg$^{1/3}$

Span (mm)	Impulsive regime		Quasi static dynamic regime	
	d_{eff} (mm)	Shear links (mm^2/m^2)	d_{eff} (mm)	Shear links (mm^2/m^2)
1,000	248	0.00	215	3,934
1,500	229	0.00	255	4,647
2,000	245	0.00	280	3,203
3,000	265	0.00	350	2,878

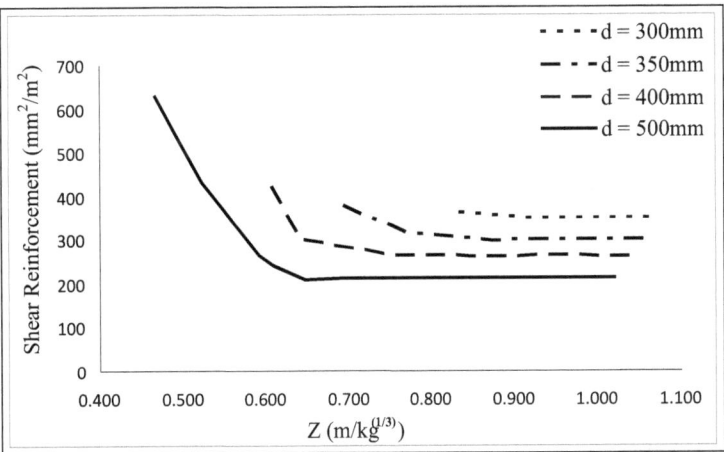

Figure 3.19: Shear reinforcement for span 3,000 mm in impulsive regime

3.10 Discussion & conclusions

Most reinforced concrete structural elements designed using conventional codes and practices have a certain blast resistant capacity. Knowing the blast resistant capacity, behavior and response of a conventional design in a blast loading environment helps in improving the design. Blast resistant capacities of structural elements can be adjudged using design envelopes. In this research, design envelopes were developed for a range of reinforced concrete cantilevered slabs.

In general, conventional designs can be improved so that they can withstand blast loading. The magnitude of the blast expected near a structure should be determined by analyzing various factors related to the purpose of the structure, its location, safety requirements, importance of the structure and social and political climate of the region. With respect to structural design, if the requirement is to protect the structure from collapse, the structure should at least satisfy the impulsive design criteria while if a much safer structure is needed, quasi static dynamic design should be used.

The results of this study show that the improvements needed to move a conventional design towards the impulsive regime are minor and easily achievable. Pushing a conventional design towards quasi static and dynamic regime, however, require many changes which make quasi static and dynamic designs costly. In order to minimize the cost, the designer has to first investigate the level of protection required for the structure and then select between conventional, impulsive and quasi static and dynamic criteria. Designing various parts of a structure with appropriate design criteria combining conventional, impulsive and quasi static dynamic designs together will produce a safe and economical structure.

From numerical estimations, it was observed that increasing the thickness of slabs (effective depth) is more effective in improving the blast resistance of cantilevers than increasing reinforcement. It was seen that there is a limit to the amount of steel needed for a cantilevered slab to be in the impulsive regime. If steel is increased (without increasing the effective depth), the slab will move away from the impulsive regime. It was also observed that any cantilevered slab can be upgraded to the quasi static and dynamic regime. The necessary improvements should be decided on based on the expected magnitude of the blast.

Most of the conventional cantilever slab designs can be analyzed using the envelopes developed in this research as they cover most practical spans and thicknesses. It should be noted that the concrete grade used in these designs was 35 N/mm^2 with steel of yield strength 460 N/mm^2. Even though this research gives envelopes only for cantilever slabs, the methodology presented can be used for any structural element.

REFERENCES

1. Agnew E, Marjanishvili S, Gallant S, 2007, Concrete detailing for blasts, Structures Magazene, Feb-2007, pp. 26-28.
2. Ayvazyan H, Dede M, Dobbs N, Whiteny M, Bowles P, Baket W, Caltagirone J.P, 1986, Structures to resist the effects of accidental explosions, Vol-02, Armamen Center, USA.
3. Beshara F.B.A, 1994, Modeling of Blast Loading on Above Ground Structures – 1, General Phenomenology and External Blast, Computers and Structures, Vol-51, No:05, Elsevier Science Ltd, UK.
4. Bing L, Huang Z, Rong H, 2006, Deformation Controlled Design of RC Flexural Members Subjected to Blast Loading, Nanyang Technological University, Singapore.
5. BS 6399-1:1996, Loading for buildings, Part 1 code of practice for dead and imposed loads, BSI, UK, 1996.
6. BS 8110-1:1997, Structural use of concrete, Part 1 code of practice for design and construction, BSI, UK, 1997.
7. BS 8110-2:1985, Structural use of concrete, Part 2 code of practice for special circumstances, BSI, UK, 1985.
8. BS 8110-3:1985, Structural use of concrete, Part 3 charts for single reinforced beams, doubly reinforced beams and rectangular columns, BSI, UK, 1985.
9. BS 8660:2000, Specification for scheduling, dimensioning, bending and cutting of steel reinforcement for concrete, BSI, UK, 2002.
10. BSEN 1990:2002, Euro code – basis of structural designs, BSI, UK, 2004.
11. Bulmash G, Kingery, C.N, 1984, Air Blast Parameters from TNT Spherical Air Burst and Hemispherical Surface Burst. Technical Report, ARBRL-TR-02555, U.S. Army Armament Research and Development Center, USA.
12. Christensen S.O, Omang M, 2009, Height of burst explosions: a comprehensive study of numerical and experimental results, Norwegian Defense State Agency, Oslo, Norway.
13. Cormie D, Mays G & Smith P, 2009, Blast Effects on Buildings 2nd edition, Thomas Telford, London, UK.
14. Dharaneepathy M.V, Rao N.K, Santhakumar A.R, 1995, Critical Distance for Blast Resistant Design, Computers and Structures, Vol-54, No: 04, Elsevier Science Ltd, UK.
15. Elliott C.L, Mays G.C, Smith P.D, 1994, The Protection of Buildings against Terrorism and Disorder, Engrs Standards and Bldgs, Paper 9930. Belfast, Ireland.
16. EN1991-1-7:2006, Eurocode 1: Actions on structures, Part 1.7: General actions, Accidental actions, CEN, Brussels, Belgium, 2006.
17. EN1992-1-1:2004, Eurocode 2: Design of concrete structures, Part 1.1: General rules and rules for buildings, CEN, Brussels, Belgium, 2004.
18. Lam N, Mendis P, Ngo T, 2004, Response Spectrum Solutions for Blast loading, Electronic Journal of Structural Engineering, The University of Melbourne, Australia.

19. Mao Y, 2006, Blast Effects and Mitigation for Conventional Buildings, The 10th Asia Pacific Conference on structural engineering and construction - August 3 – 5, 2006, Bangkok, Thailand.

20. Mc Cann D. M, Smith S.J, 2007, Blast resistant design of reinforced concrete structures, Structures Magazene, Apr-2007, pp. 22-26.

21. Remennikov A.M, 2009, State of the Art of Explosive Loads Characterization, University of Wollongong, Australia.

22. Remennikov A.M, 2003, A Review of Methods for Predicting Bomb Blast Effects on Buildings, Journal of Battlefield Technology, Vol: 06, No:03, University of Wollongong, Australia.

23. Rosenthal M.F, Morlock G.L, 1987, Blasting Guidance Manual, US Department of the Interior, USA.

24. Rouzsky N, 1988, Blast-resistant Control Buildings, Structural Safety, 5 (1988) 253-266, Elsevier Science Publishers, B.V., Amsterdam, The Netherlands.

25. Schmidt P.E, 2003, Structural Design for External Terrorist Bomb Attacks, Structures Magazine, Page 14/15, Burns & McDonnell, Kansas City, USA.

26. Swisdak M.M, 2009, Simplified Kingery Air blast Calculations, Naval Surface Warfare Center, USA.

27. UFC-03-340-01, 2002, Design and analysis of hardened structures to conventional weapons effects, Department of Defense, USA, 2008.

28. UFC-03-340-02, 2008, Structures to resist the effects of accidental explosions, Department of Defense, USA, 2008.

29. UFC-04-10-01, 2007, DoD minimum antiterrorism standards for buildings, Department of Defense, USA, 2007.

ANNEXES

List of Annexes

Annex 1.1:

Kingery and Bulmash's Solution for Hemispherical Surface Blasts

(a) Graphical illustration $(0.067 \text{ m/kg}^{1/3} < Z < 40 \text{ m/kg}^{1/3})$

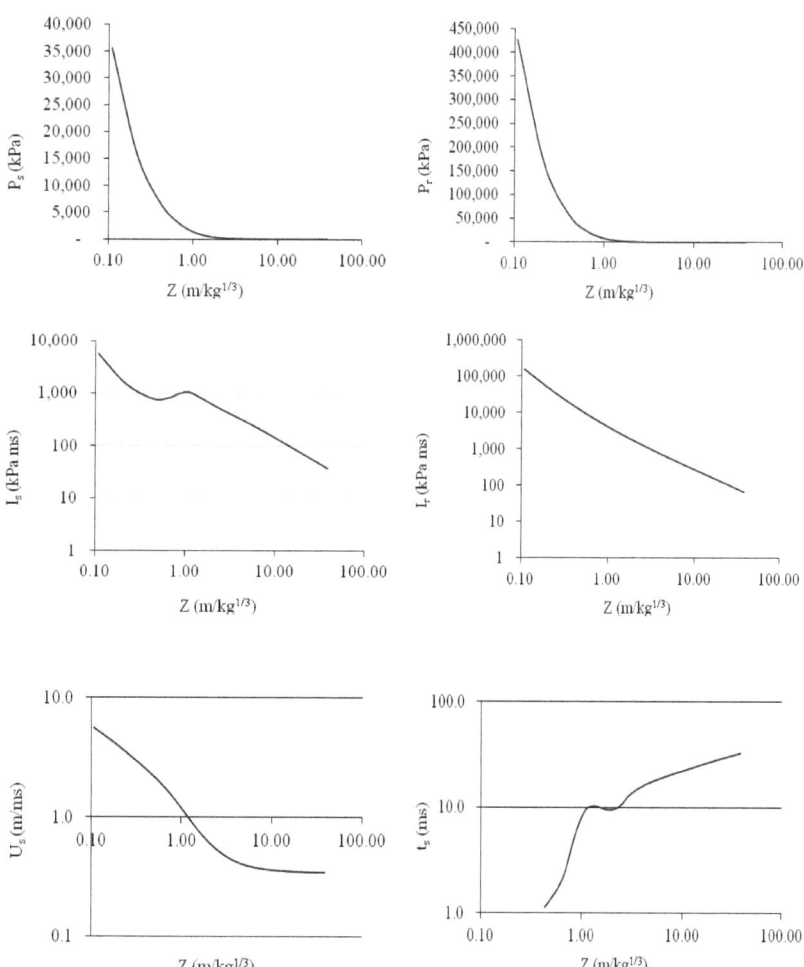

(b) Blast parameters in tabular form ($0.067 \text{ m/kg}^{1/3} < Z < 40 \text{ m/kg}^{1/3}$)

R (m)	W (kg)	Z (m/kg$^{1/3}$)	P_s (kPa)	P_r (kPa)	I_s (kPa ms)	I_r (kPa ms)	U_s (m/ms)	t_s (ms)
0.5	100	0.11	35,653	426,906	5,599	154,297	5.7	
1.0	100	0.22	15,737	165,803	1,523	42,880	3.8	
2.0	100	0.43	6,120	51,998	790	13,800	2.4	1.1
3.0	100	0.65	3,173	23,490	816	7,542	1.8	2.1
4.0	100	0.86	1,843	12,002	996	5,029	1.4	5.6
5.0	100	1.08	1,157	6,645	1,064	3,717	1.1	9.3
6.0	100	1.29	773	3,942	938	2,925	0.9	10.3
7.0	100	1.51	544	2,488	818	2,400	0.8	10.0
8.0	100	1.72	400	1,659	720	2,029	0.7	9.6
9.0	100	1.94	305	1,160	643	1,754	0.6	9.5
10.0	100	2.15	239	845	582	1,543	0.6	9.7
11.0	100	2.37	193	639	532	1,375	0.6	10.2
12.0	100	2.59	159	498	492	1,240	0.5	11.1
13.0	100	2.80	134	398	458	1,128	0.5	12.4
14.0	100	3.02	114	326	428	1,035	0.5	13.2
15.0	100	3.23	99	272	403	955	0.5	13.9
16.0	100	3.45	87	231	381	887	0.4	14.6
17.0	100	3.66	77	199	362	827	0.4	15.1
18.0	100	3.88	69	174	344	775	0.4	15.6
19.0	100	4.09	62	154	328	729	0.4	16.1
20.0	100	4.31	56	137	314	688	0.4	16.5
21.0	100	4.53	52	124	301	651	0.4	16.9
22.0	100	4.74	47	112	288	619	0.4	17.2
23.0	100	4.96	44	103	277	589	0.4	17.5
24.0	100	5.17	41	94	267	562	0.4	17.8
25.0	100	5.39	38	87	257	537	0.4	18.1
30.0	100	6.46	28	63	218	440	0.4	19.3
35.0	100	7.54	22	48	189	372	0.4	20.3
40.0	100	8.62	18	39	166	323	0.4	21.2
50.0	100	10.77	13	28	134	255	0.4	22.8
60.0	100	12.93	11	22	112	210	0.4	24.1
70.0	100	15.08	9	18	97	179	0.4	25.3
80.0	100	17.24	7	15	85	155	0.4	26.4
90.0	100	19.39	6	13	76	137	0.3	27.3
100.0	100	21.55	6	11	69	123	0.3	28.1
110.0	100	23.70	5	10	62	111	0.3	28.9
120.0	100	25.86	4	9	57	102	0.3	29.5
130.0	100	28.01	4	8	53	94	0.3	30.2
140.0	100	30.17	4	7	49	87	0.3	30.7
150.0	100	32.32	3	7	46	81	0.3	31.3
180.0	100	38.79	2	5	38	66	0.3	32.9

Annex 1.2:
Reflection Coefficient (C$_{ra}$), Ayvazyan et al. (1986), Cormie et al., (2009)

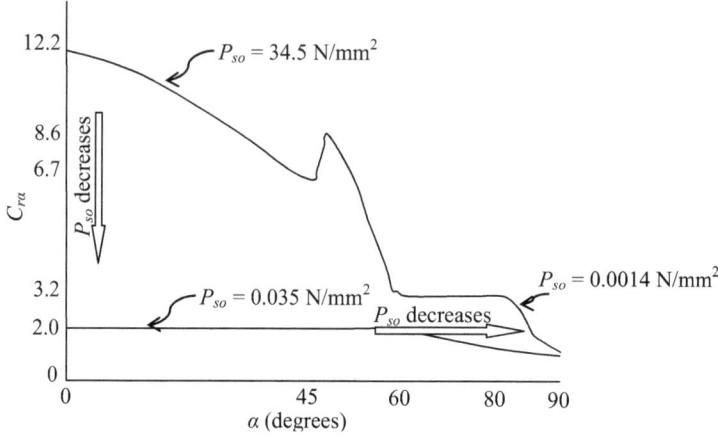

Annex 1.3:
Equivelent SDOF factors for cantilevers with uniformly distributed loads

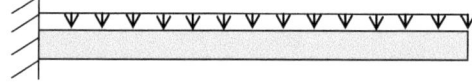

Length of the cantilever = L

Uniformly distributed load per unit length = p

Uniformly distributed load $F = pL$

Uniformly distributed mass per unit length = m

Uniformly distributed mass $M = mL$

	Elastic	Plastic
Strain range	0.4	0.5
Load factor K_L	0.26	0.33
Mass factor K_M	0.26	0.33
Load mass factor K_{LM}	0.65	0.66
Maximum resistance R_m	$2Mp/L$	$2M_p/L$
Stiffness k	$8EI/L^3$	0
Support shear V_s	R_m	R_m
Ultimate shear stress, v	$R_m(1/de - 1/L)$	$R_m(1/de - 1/L)$
Dynamic reaction	$0.69R + 0.31F$	$0.75R_m + 0.25F$

M_p is the ultimate hogging moment capacity at support, de is the distance from face of the support and v is the ultimate shear at distance de.

Annex 1.4:
Cofficient (F) for second moment of area of reinforced concrete sections,
Cormie et al., (2009)

(a) **Data used for design calculations for cracked sections with only tension**
reinforcement

$A_s/(bd)$	$\alpha_e = E_s/E_c$				
	5	6	8	10	12
0.000	0.000	0.000	0.000	0.000	0.000
0.002	0.008	0.010	0.012	0.015	0.017
0.004	0.015	0.018	0.023	0.027	0.032
0.008	0.027	0.032	0.040	0.047	0.055
0.012	0.038	0.045	0.055	0.065	0.074
0.016	0.048	0.055	0.068	0.078	0.088
0.020	0.056	0.064	0.079	0.090	0.102
			$F = I_c/(bd^3)$		

(b) **Data used for design calculations for cracked sections with equal reinforcement**
on opposite faces for tension and compression

$A_s/(bd)$	$\alpha_e = E_s/E_c$				
	5	6	8	10	12
0.000	0.000	0.000	0.000	0.000	0.000
0.002	0.008	0.010	0.013	0.016	0.018
0.004	0.015	0.018	0.023	0.027	0.031
0.008	0.028	0.032	0.041	0.049	0.057
0.012	0.039	0.046	0.057	0.069	0.080
0.016	0.049	0.057	0.073	0.087	0.101
0.020	0.060	0.068	0.086	0.103	0.120
			$F = I_c/(bd^3)$		

Annex 1.5:
Maximum deflection of elasto-plastic, SDOF system for trianguler load, Cormie et al., (2009)

The data used for x_m/x_e in design calculations

t_d/T	r_u/P								
	0.1	0.2	0.4	0.6	0.8	1	1.2	1.5	2
0.1	5	1.6	0.7	0.5	0.4	0.35	0.3	0.2	0.16
0.2	18	5	1.5	0.9	0.7	0.6	0.5	0.4	0.33
0.5	100	30	7	3	1.6	1.2	0.9	0.75	0.6
1		100	20	7	3.5	1.8	1.3	1	0.8
2			70	20	7	3.3	1.8	1.2	0.85
5				90	19	4.9	2.3	1.4	0.95
10					65	7.5	2.9	1.5	1
20						11	3	1.5	1

x_m/x_e

Annex 1.6:
Maximum response time of elasto-plastic, SDOF system for trianguler load, Cormie et al., (2009)

The data used for t_m/t_d in design calculations

t_d/T	r_u/p								
	0.1	0.2	0.4	0.6	0.8	1	1.2	1.5	2
0.1	6	4	3.3	3.3	3.3	3.3	3.3	3.3	3.3
0.2	5.5	3.3	1.9	1.7	1.6	1.6	1.6	1.6	1.6
0.5	5	3	1.6	1.1	0.9	0.8	0.8	0.8	0.8
1	5	2.9	1.5	1	0.7	0.6	0.5	0.45	0.45
2	5	2.9	1.5	0.9	0.6	0.4	0.3	0.3	0.25
5	5	2.8	1.4	0.9	0.5	0.3	0.15	0.11	0.1
10	4.9	2.8	1.4	0.9	0.4	0.19	0.09	0.06	0.05
15	4.9	2.7	1.4	0.9	0.4	0.17	0.07	0.04	0.035

t_m/t_d

Annex 2.1:

Conventional design of cantilever slabs

Description / Parameter	Span (mm)			
	1,000	1,500	2,000	3,000
L_x (mm)	1,000.00	1,500.00	2,000.00	3,000.00
b (mm)	1,000.00	1,000.00	1,000.00	1,000.00
q_k (kN/m^2)	5.00	5.00	5.00	5.00
d_{eff} (mm)	110.00	165.00	220.00	330.00
D (mm)	140.00	196.00	251.00	361.00
g_k (kN/m^2)	4.40	5.70	7.00	9.70
n (kN/m^2)	14.10	15.99	17.83	21.53
M_d (kNm)	7.05	17.98	35.65	96.88
$M/(bd^2)$	0.58	0.66	0.74	0.89
Span/d_{eff}	10.60	10.20	9.90	9.40
Span/$d_{eff\,pro}$	9.10	9.10	9.10	9.10
Result	ok	ok	ok	ok
K'	0.156	0.156	0.156	0.156
K	0.017	0.019	0.021	0.025
Result	ok	ok	ok	ok
z (mm)	80.00	120.00	160.00	238.00
$0.95d$ (mm)	105.00	157.00	209.00	314.00
Result	ok	ok	ok	ok
A_s (mm^2)	200.55	342.11	510.58	930.81
A_s min (mm^2)	182.00	255.00	326.00	469.00
Result	ok	ok	ok	ok
V (kN/m)	14.10	23.98	35.67	64.59
v_c (N/mm^2)	0.55	0.46	0.46	0.42
Result	ok	ok	ok	ok
$V/(u_o d)$ (N/mm^2)	0.040	0.030	0.020	0.010
$V/(ud)$ (N/mm^2)	0.018	0.010	0.006	0.003
Result	ok	ok	ok	ok
A_c (mm^2)	110,000	165,000	220,000	330,000
A_s/A_c	0.0018	0.0021	0.0023	0.0028

Annex 2.2:
Impulsive design of a span 1,000 mm cantilever slab

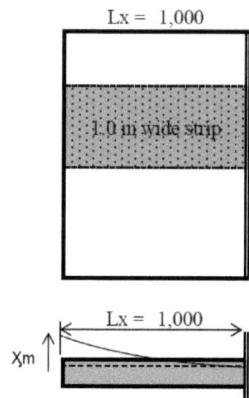

Lx = 1,000

1.0 m wide strip

b =	1,000	mm
Lx =	1,000	mm

Lx = 1,000

Xm

For a cantilever, dead and imposed loads act on the slab but blast loading acts from under side. The resultant bending moment and shear force reduce or become negative depending on the magnitude of blast. In this calculation, the magnitude of blast loading is considerably higher and therefore the effect of dead and imposed loads are not added to bending moments and shear forces.

Steel to Concrete area ratio,   As/bd =   0.0010
ρ (initial),

Determine the predicted weight of explosives & suitable stand off distance

$$Z = R/W^{(1/3)}$$

Designing for 100kg of TNT at ground level at stand off distance of 4m.

$$Z = 2.16 \ m/kg^{1/3}$$

Calculate blast parameters
Using Kingerly and Bulmash equations or directly from Table 2.2,

Positive phase duration, t_s (t_d) =	9.70	ms
Reflected impulse, Ir =	1,543	kPa ms (kN ms / m^2)
Reflected over pressure Pr =	845	kPa (kN / m^2)

Define the resistance, deflection function in terms of Rm

Ultimate resistance of the element, Rm = $f(M_{Rd}.L)$

For cantilever, from Annex-1.3, $Rm = 2Mp/L$ where $Mp = M_{Rd}$
where Mp is the ultimate hogging moment capacity at support.

Determine the required protection category

If protection category is 2, (Type 2 section where concrete is crushing and no capacity to take moments),

must be double reinforced,
$\theta > 2°$
implusive loading

Calculate MRd and Rm as a function of z

For type 2 section,　　　　　　　　　$M_{Rd} = As \, f_{yd.dyn} \, z/b$

　　　　　　　　　　　　　　　　　　$\rho_1 = As/(bz)$

Assuming $\rho_1 = \rho$　　　　　　　　$\rho_1 = \quad 0.001$

Therefore,　　　　　　　　　　　　$M_{Rd} = \rho_1 \, f_{yd.dyn} \, z^2$

From Annex-1.3 For cantilever slab, $Rm = 2Mp/L$

　　　　　　Therefore $Rm = (2/L)(\rho_1 \, f_{yd.dyn} \, z^2)$
　　　　　　　　　　　　where z is the lever arm.

For reinforcement,

$f_{yd.dyn} =$	1.20	f_{yk}　(Table 2.4 for $f_{yd.dyn}$)
$f_{yk} =$	460.00	N/mm^2
$f_{yd.dyn} =$	552.00	N/mm^2
$f_{ck} =$	35.00	N/mm^2

Therefore,　　　　　　　　　　　$Rm = \quad \underline{1,104.00} \; z^2$　　　kN per meter width.

Estimate Deflection at elastic limit, Xe

θ to be determined between 2 ~ 5 Deg., according to links etc.

Therefore,　　　　　　Assume, $\theta = \quad 4$ Deg.

Therefore,　　　Total deflection, $Xm = \quad$ Lx tanθ
　　　　　　　　　　　　　　　　　69.94 mm

The equievalent elastic stiffness,　　$ke = f[E,I,L]$

From Annex-1.3, for cantilever,　　$ke = \quad 8EI/L^3$

$$Ec = 3.40E+10 \text{ Pa}$$
$$Es = 2.00E+11 \text{ Pa}$$

From Annex-1.4, for cracked section with equal reinforcement in opposite faces assuming cover is lost on both faces,

$$\alpha = Es/Ec$$
$$5.88$$

$$\rho = 0.0010$$

Therefore, $\quad F = 0.0050$

Therefore, $\quad Ic = Fbd^3$
$$0.0050 \; bz^3$$

$$ke = 1.36E+06 \; z^3 \qquad \text{kN/m width.}$$

Therefore, Deflection at elastic limit, $Xe = Rm/ke$

$$8.12E-04 \; /z \qquad m$$

Determin load mass factor KLM as a function of z

From Annex-1.3, Load mass factor, $K_{LM} = 0.66$

Density of concrete, $\rho_c = 2,500.00 \text{ kg/m}^3$

Mass of the element, $M = \rho_c \; z \; b \; L$

$$2,500.00 \; z \qquad \text{kg/m width.}$$

Substitute values in the basic Implusive equation to solve for z

$$I^2 A^2/(2K_{LM}M) = Rm(Xm - Xe/2)$$

$$I = 1,542.63 \text{ kNms/m}^2$$

$$A = L \; b$$
$$1.00 \text{ m}^2$$

$$z = 0.213 \text{ m}$$

$$z = 0.213 \text{ (Solution for z)}$$

Determin the steel requirement

$$As = \rho_1 \; z \; b$$
$$213.09 \text{ mm}^2$$

Diameter of bar, $\quad D = 12.00 \text{ mm}$

Reinforcement cover, $\quad C = 25.00 \text{ mm}$

Therefore, $\quad h = 275.09 \text{ mm}$

Area of one bar, $\quad = 113.04 \text{ mm}^2$

Number of bars required, $\quad = 1.89 \text{ Nos}$

No of bars provided, $\quad = 1.89 \text{ Nos}$

Bar spacing $\quad Sv = 530.00 \text{ mm}$

Therefore, As, Provided $= 213.09 \text{ mm}^2$

Check the sufficiency (impulsive)

$$Rm = 50.13 \text{ kN per meter width}$$

Therefore, unit resistance, $ru = Rm/Lx$

		50.13	kPa per meter width
time at maximum deflection, tm =		I_r/r_u	
		30.77	ms
	tm/ts =	3.17	
For implusive regiem,	tm/ts >	3.00	
	Therefore considered section is	OK	

Implusive loading design is   valid

Flextural capacity is   adequate

BS 8110 / Part-I / 1997

Shear Reinforcement:

From Annex-1.3 for cantilever,

Maximum Shear Force = Rm   50.13 kN

Ultimate design shear, V_{ED} =   Rm[1/d-1/L]*d

V_{ED} =   39.45 kN

Shear Stress, v = V_{ED}/(bvd) (d = z)

v =   0.19 N/mm^2

$100As/b_vd$ =   0.10

d_{eff} = z =   212.6 mm

Therefore vc =   0.40 N/mm^2

$0.5v_c$	0.20	N/mm^2
$0.5v_c + 0.4$	0.80	N/mm^2
$0.8 \sqrt{fcu}$	4.31	N/mm^2	(for shear, fcd.dyn = 0.83fck, therefore check for
5 N/mm^2	5.00	N/mm^2	$0.8\sqrt{0.83 fcu}$)

Case-1	$v < v_c$	OK
Case-2	$v_c < v < v_c + 0.4$	CHECK
Case-3	$v_c + 0.4 < v < 0.8 \sqrt{fcu}$ or 5N/mm^2	CHECK

If case 01: No need shear links.

If case 02: Need shear links;   Asv = 0.4 b Sv/(0.95 fyv)

Min spacing, 0.75d =	159.43	mm		
Assiumed spacing, Sv =	175.00	mm		
fy, for links =	250.00	N/mm2		
Therefore Asv =	294.74	mm2		
Reinforcement for Links, dia. =	10.00	mm		
Nomber of links at one plane =	2.00	Nos		
Area of steel in one stirrup =	314.00	mm2	> Asv	OK
Provide dia 10.00 um links @	175.00	mm spacing everywhere		

If case 03: Need shear links &/or bent up bar (shear link to be calculated as follows)

$$Asv = b \, Sv \, (v\text{-}vc)/(0.95fyv)$$

Min spacing, 0.75d =	159.43	mm
Assiumed spacing, Sv =	150.00	mm
fy, for links =	250.00	N/mm2
Therefore Asv =	(132.27)	mm2
Reinforcement for Links, dia. =	10.00	mm
Nomber of links at one plane =	4.00	Nos
Area of steel in one stirrup =	628.00	mm2 > Asv OK
Provide dia 10.00 um links @	150.00	mm spacing everywhere

Check for Shear:

v =	0.19	N/mm^2
0.8 √ fcu =	4.31	N/mm^2
Max. =	5.00	N/mm^2
v < 0.8 √ fcu or 5N/mm^2		OK

Results:

Parameter	Value	Unit
fy	460	N/mm^2
fcu	35	N/mm^2
d	248	mm
Ac = bd	248,092	mm^2/m
As	213	mm^2/m
As + A's	426	mm^2/m
$\rho_t = (As+As')/Ac$	0.17	%
As,transverse	Minimum	mm^2/m
As, shear per stirriup	N/R	mm^2/strriup
As, shear	N/R	mm^2/m length

Annex 2.3:
Quasi static and dynamic design of a span 1,000 mm cantilever slab

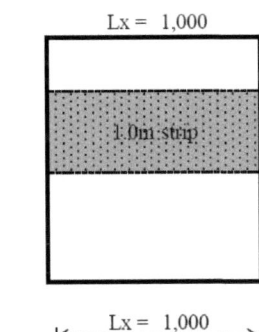

$$b = \quad 1{,}000 \quad mm$$
$$Lx = \quad 1{,}000 \quad mm$$

Lx = 1,000

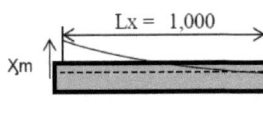

For a cantilever, dead and imposed loads act on the slab but blast loading acts from under side. The resultant bending moment and shear force reduce or become negative depending on the magnitude of blast. In this calculation, the magnitude of blast loading is considerably higher and therefore the effect of dead and imposed loads are not added to bending moments and shear forces.

$$\text{Thickness, } h = \quad 250 \quad mm$$
$$\text{Concrete Area, } Ac = \quad 2.50E+05 \ mm^2$$

Assuming ρ, $As/bd =$ 0.0080

Determine the predicted weight of explosives & suitable stand off distance.
$$Z = \quad R/W^{1/3}$$

Designing for 100kg of TNT at ground level at stand off distance of 4m.
$$Z = \quad 2.16 \ m/kg^{1/3}$$

Calculate blast parameters
Using Kingerly and Bulmash equations or directly from Table 1,

Positive phase duration, t_s (t_d) =	9.70 ms	
Reflected impluse, Ir =	1,543 kPa ms	(kN ms / m²)
Reflected over pressure Pr =	845 kPa	(kN / m²)

Define the resistance, deflection function in terms of Rm

Ultimate resistance of the element, $Rm = f(M_{Rd}, L)$
For cantilever, from Annex-1.2, $Rm = 2Mp/L$ where $Mp = M_{Rd}$

where Mp is the ultimate hogging moment

69

Determine the required protection category
If protection category is 1, (Type 1 section where concrete is effective on both surcfaces),
can be single or double reinforced,
θ in the range of $0 \sim 2°$
dynamic or quasic-static

Calculate MRd and Rm and hence ru/Pr
For type 1 section, $\qquad M_{Rd} = As\, f_{yd,dyn}\, (d\text{-}0.5\lambda x)/b$

From Annex-1.2 For cantilever slab, $Rm = 2Mp/L$

$$x = As\, f_{yd.dyn}/(b\, \eta\lambda\, f_{cd,dyn})$$
where, x is the depth from compression
phase to nutral axis,
λ is the proposion of x over which rectanguler
stress distribution of magnitude ηfcd acts,
η is the factor on fcd giving the magnitude of
rectanguler stress distribution.

Simplyfied equations for fck < 50MPa, (λ=0.8, η=1.0)
$$M_{Rd} = 1.2\, As\, f_{yk}\, (d\text{-}0.4x)/b \qquad \text{(Table 4.2 for fyd,dyn)}$$
$$x = As\, f_{yk}/(0.59\, b\, f_{ck}) \qquad \text{(Table 4.2 for fcd,dyn)}$$

As/Ac =	0.0080	
d =	215.00	mm
h =	250.00	mm
f_{yk} =	460.00	N/mm*
f_{ck} =	35.00	N/mm*

As = A's =		1,720.00	mm^2
Diameter of bar,	D =	20.00	mm
Reinforcement cover,	C =	25.00	mm
Area of one bar,	=	314.00	mm^2
Number of bars required,	=	5.48	Nos
No of bars provided,	=	5.48	Nos

| | Bar spacing | $Sv =$ | 183.00 | mm |
| | | As, Provided $=$ | 1,720.00 | mm^2 |

Therefore,

	$x =$	38.31	mm
	$z =$	$d-0.4x$	
	$z =$	199.67	mm
	$M_{Rd} =$	189.58	kNm per 1m width.
	$Rm =$	379.16	kN per 1m width.

Therefore, unit resistance, $r_u =$ Rm/Lx

379.16 kN/m per 1m width.

Therefore, $r_u/Pr =$ 0.4484

Calculate the maximum allowable deflection

\# θ to be determined between 0 ~ 2 Deg., according to links etc.

\# In the absense of links in protection category1, $\theta = 1$ Deg.

\# Maximum value for protection category1, $\theta = 2$ Deg.

Therefore, Assume, $\theta =$ 2 Deg.

Therefore, Total deflection, Xm all= Lx tanθ

34.93 mm

Calculate the deflection at elastic limit

From Annex-1.2, for cantilever, $ke =$ 8Ec I/L^3

$Ec =$ 3.40E+10

$Es =$ 2.00E+11

From Annex-1.3, for cracked section with tension reinforcement assuming cover is lost only on the tensile face,

	$\alpha e =$	Ec/Ec
		5.88
	$\rho =$	0.0080

Therefore, $F =$ 0.0330

Therefore, $Ic =$ Fbd^3

3.28E-04

$ke =$ 8.92E+07 N/m

Therefore, Deflection at elastic limit, $Xe =$ Rm/ke

4.25 mm

Calculate the natural period of the element and hense ts/T

Natural period of the element, $T = 2\pi\sqrt{(K_{LM} M/ke)}$

From Annex-1.2, Load mass factor, $K_{LM} =$ 0.65

Density of concrete, $\rho c =$ 2,500.00 kg/m^3

Mass of the element, $M =$ $\rho c\ h\ b\ L$

625.00 kg/m width.

Therefore, $T =$ 13.41 ms

$t_s/1 =$ 0.7236

Calculate the maximum deflection and check with the deflection at the limit of θ

From Annex-1.5, maximum deflection of elasto-plastic SDOF system for trianguler blast load,

$$Xm/Xe = \quad 7.50$$

Therefore,
$$Xm = \quad 31.88 \quad mm$$
$$Xm \text{ (assumed)} = \quad 34.93 \quad mm$$

Therefore considered section is   OK

Check whether the response of the structure is within quasi static - dynamic regiem or not

From Annex-1.6, maximum response time of elasto-plastic SDOF system for trianguler blast load,

$$t_m/t_d = \quad 1.40$$

For quasi static - dynamic regiem,
$$t_m/t_d < \quad 3.00$$

Therefore considered section is   OK

Quasistatic - dynamic loading design is   valid

Flextural capacity is   adequate

BS 8110 / Part-I / 1997

Shear Reinforcement:

From Annex-1.2 for cantilever,

Maximum Shear Force $Rm = \quad 379.16 \quad kN$

Ultimate design shear $V_{ED} = \quad Rm[1/d-1/kN$

$$V_{ED} = \quad 297.64 \quad kN$$

Shear Stress, $v = \quad V/(bvd)$

$$v = \quad 1.38 \quad N/mm^2$$

$100As/bvd = \quad 0.80$

$d_{eff} = \quad 215.0 \quad mm$

Therefore $vc = \quad 0.67 \quad N/mm^2$

$0.5vc \quad\quad\quad 0.34 \quad N/mm^2$

$0.5vc + 0.4 \quad\quad 1.07 \quad N/mm^2$

$0.8 \sqrt{fcu} \quad\quad 4.31 \quad N/mm^2 \quad$ (for shear, $f_{cd.dyn} = 0.83f_{ck}$,

$5 \ N/mm^2 \quad\quad 5.00 \quad N/mm^2 \quad$ therefore check for $0.8\sqrt{0.83\ f_{cu}}$)

Case-1 \quad\quad $v < vc$ \quad\quad\quad\quad\quad\quad\quad\quad\quad\quad\quad FALSE

Case-2 \quad\quad $vc < v < vc + 0.4$ \quad\quad\quad\quad\quad\quad FALSE

Case-3 \quad\quad $vc + 0.4 < v < 0.8 \sqrt{fcu} \text{ or } 5N/mm^2$ \quad CHECK

If case 01 No need shear links.

If case 02 Need shear links; \quad\quad\quad $Asv = 0.4 \ b \ Sv/(0.95 \ f_{yv})$

Min spacing, $0.75d = \quad 161.25 \quad mm$

Assiumed spacing, $Sv = \quad 175.00 \quad mm$

fy, for links $= \quad 250.00 \quad N/mm2$

Therefore $Asv = \quad 294.74 \quad mm2$

Reinforcement for Links, dia. $= \quad 10.00 \quad mm$

Nomber of links at one plane $= \quad 2.00 \quad Nos$

Area of steel in one stirrup $= \quad 314.00 \quad mm2 \quad > Asv \quad OK$

Provide dia.   10.00 um links @   175.00 mm spacing everywhere

If case 03 Need shear links &/or bent up bars; (shear link to be calculated as follows)

$$Asv = b \ Sv \ (v\text{-}vc)/(0.95fyv)$$

Min spacing, 0.75d =	161.25	mm
Assiumed spacing, Sv =	175.00	mm
fy, for links =	250.00	N/mm^2
Therefore Asv =	526.37	mm^2
Reinforcement for Links, dia. =	10.00	mm
Number of links at one plane =	6.00	Nos
Area of steel in one stirrup =	942.00	mm^2 > Asv OK
Provide dia. 10.00 um links @	175.00	mm spacing everywhere

Check for Shear:

v =	1.38	N/mm^2
0.8 $\sqrt{}$ fcu =	4.31	N/mm^2
Max. =	5.00	N/mm^2
v < 0.8 $\sqrt{}$ fcu or 5N/mm^2		OK

Results:

Parameter	Value	Unit
fy	460	N/mm^2
fcu	35	N/mm^2
d	215	mm
Ac	215,000	mm^2/m
As	1,720	mm^2/m
As + A's	3,440	mm^2/m
ρ = As/Ac	1.60	%
As,transverse	Min	mm^2/m
As, shear	526	mm^2/strriup
As, shear	3,008	mm^2/strriup

73

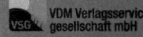

Printed by Books on Demand GmbH, Norderstedt / Germany